MECHANICAL ENGINEERING THEORY AND APPLICATIONS

NATURAL VENTILATION

STRATEGIES, HEALTH IMPLICATIONS AND IMPACTS ON THE ENVIRONMENT

MECHANICAL ENGINEERING THEORY AND APPLICATIONS

Additional books in this series can be found on Nova's website under the Series tab.

Additional e-books in this series can be found on Nova's website under the e-book tab.

MECHANICAL ENGINEERING THEORY AND APPLICATIONS

NATURAL VENTILATION

STRATEGIES, HEALTH IMPLICATIONS AND IMPACTS ON THE ENVIRONMENT

OSCAR T. HAYNES
EDITOR

nova publishers
New York

Library of Congress Cataloging-in-Publication Data
Natural ventilation : strategies, health implications and impacts on the environment / editor, Oscar T. Haynes.
pages cm. -- (Mechanical engineering theory and applications)
Includes index.
ISBN 978-1-63482-414-9 (hardcover)
1. Natural ventilation. I. Haynes, Oscar T.
TH7674.N36 2014
697.9'2--dc23

2015007253

Published by Nova Science Publishers, Inc. † New York

CONTENTS

PREFACE

This book beings with an introduction into natural ventilation applications in buildings. It defines the concept of natural ventilation in buildings, and compares natural and mechanical ventilation strategies. It continues by discussing commercial win towers, which is a passive ventilation technology based on traditional vernacular wind towers of the Middle East, particularly in climates of hot, arid conditions. Furthermore, this book evaluates indoor air quality in school buildings; thermal comfort conditions in spaces equpped with radiant floor systems; and differential ventilation strategies in building spaces.

Chapter 1 – Natural ventilation is a sustainable design strategy. However, its utilisation in buildings requires understanding of many relevant aspects including its driving forces and strategies. Natural ventilation is of a great environmental advantage, compared to the mechanical ventilation strategy. This has encouraged several researches on natural ventilation implementation in buildings, considering its main objective of providing buildings with the required air quality and quantity.

A review of human thermal comfort revealed that natural ventilation has a high potential for improving thermal comfort in hot climates, implementing wind-induced ventilation, and in cold climates, implementing temperature-induced ventilation. Within this context, natural ventilation can be used in several strategies and techniques for passive heating or cooling. Natural ventilation affects different heat transfer mechanisms between human body and its ambient environment. This mostly occurs in the form of sensible heat, by convection and radiation, and the form of latent heat, by evaporation of moisture on the skin.

Chapter 2 – The energy crisis drives the design of the workplace towards passive systems, such as natural ventilation. The design of the ventilation system influences the energy demand of the building and comfort of the user. In order to improve the design, optimise the energy performance, predict and improve users' comfort, the application of accurate computational modelling and analysis techniques are essential. This study reviews the advances in thermal comfort modelling and energy analysis of the workplace using Computational Fluid Dynamics (CFD) and Building Thermal Modelling (BTM). This is followed by a case study of CFD-BTM analysis of energy and thermal comfort of a practice example of a naturally ventilated office. This is a three storey building with an open plan layout in the UK and the office area of 3000 m^2, including 375 workstations. It was awarded by the British Council for Offices, received an 'excellent' BREEAM rating and 10 out of 10 Environmental Performance Indicator (EPI) rating. The office is naturally ventilated with manually and mechanically operated windows and a stack effect through the vents above the atrium. Furthermore, environmental measurements and comfort surveys are applied in the case study buildings.

Chapter 3 – Commercial wind towers are passive ventilation technology based on traditional vernacular wind towers of the Middle East, particularly in climates of hot, arid conditions. By manipulating pressure differences and the buoyancy effect, caused by wind driven flow and density stratification of internal air due to temperature differences, adequate levels of ventilation can be provided to buildings utilising wind towers in a low energy process. This negates the use of mechanical air-conditioning systems which require substantial amounts of energy. One current limitation of commercial wind towers which exists is the inability to condition the incoming air for thermal comfort to occupants. This relates to cooling demand in hot climates and heating and cooling demand in temperate climates. At present, commonly used methods to provide thermal comfort rely on energy intensive mechanical processes to cool or heat the incoming air which is not sustainable given pressure to reduce energy use and global greenhouse gas emissions.

Low energy methods for cooling and heating of air passing through a wind tower have not yet been comprehensively explored. A cooling system utilising the combination of heat pipes and a heat sink has been developed and incorporated into a commercial wind tower. This system has been analysed using CFD and tested experimentally in a hot climate and showed positive results in reducing the temperature of the incoming air. A heat recovery system utilising a rotary thermal wheel to transfer the thermal energy from the exhaust airstream of a wind tower has been developed and incorporated into a four-sided commercial wind tower.

Chapter 4 – School buildings constitute a great part of the non-residential building stock and it is particularly important to ensure a good indoor air quality therein. Natural ventilation is frequently the only ventilation system available in most of educational buildings. Therefore, ventilation rates are correlated with wind speed, wind direction, indoor-outdoor temperature difference and are closely related to building design. Unfortunately, this kind of ventilation is far from being reliable. The limited information available on the thermal comfort and indoor air quality, the absence of proper legislative measures and the lack of interest concerning ventilation efficiency may be some of the reasons for the non-resolution of this problem. School buildings face the same indoor air quality problems as any other building. Although the health and productivity of pupils and teachers is strongly affected by the indoor environmental quality of their school, a recent survey conducted in Portugal, revealed poor indoor air quality in all the schools under study, even for recently constructed or refurbished buildings. Another survey conducted in Portugal also reported that thermal comfort in an elementary school is barely achieved. Thermal comfort models are usually based on adult subjects, being no assurance that these models can be applied to children. Surveys conducted in school children aged 7-11 in Hampshire and Southampton, UK on thermal comfort indicate that children in a classroom have a different thermal perception compared with adults, showing that this is a sensitive subject to deal with that requires further studies, to ensure a comfortable and healthy environment for educational purposes. Another study showed that installation of ventilation systems with heat recovery is often difficult or even impossible in existing school buildings. Therefore, natural ventilation plays a crucial role in promoting the comfort and health of occupants. It has the potential to significantly save cooling energy during cooling season, but it is costly in terms of heating energy consumption during winter, but can also be beneficial from an energy perspective when free heating is available. Different studies have shown that double ventilated windows can promote natural aeration of indoor environment with warmer air than it is found outdoors. Besides, this system can promote sustainability and can be use both on old and new buildings. The studies were based on objective evaluation by monitoring indoor conditions and subjective occupant evaluation using questionnaires. Simulations using analytical methods for different boundary conditions such as temperature and pressure differences across openings were also used to determine thermal comfort and energy savings. The potential of this ventilated heat recovery system is evaluated in terms of energy savings and thermal comfort. The present chapter deals with effects of wind and temperature difference between indoor and outdoor environments on the stack ventilation performance in an average elementary school in Portugal. The results for the cold moderate climate regions of Portugal and other coldest climates show that while indoor air quality can be easily maintained, the heating potential and thermal comfort strongly depends on the natural ventilation strategy.

Chapter 5 – The aim of this chapter is to evaluate numerically the comfort conditions in spaces equipped with radiant floor system and forced ventilation system. The results of this original chapter are obtained in a virtual chamber equipped with a desk, two seats and two virtual mannequins. In this work the numerical softwares, developed by the author, are presented.

In this numerical work the thermal comfort, local thermal discomfort and air quality levels are evaluated in two occupants in a space equipped with radiant floor system and forced ventilation system. This study, developed from three coupled software, considers the buildings thermal response, computational fluids dynamics and human thermo-physiology models.

In this simulation the temperature field on the building surfaces and collector water are calculated from the buildings thermal response with complex topology, the field of air temperature and air velocity around the occupants is calculated from the computational fluids dynamics and the temperature field in the occupants and clothing is calculated from the human thermo-physiology models.

Chapter 6 – The aim of this chapter is to analyse the application of integral and differential ventilations strategies in buildings spaces in the South of Portugal. The results of this review chapter are obtained, in Mediterranean environment, in occupied spaces. In this work the integral and differential numerical strategies and the experimental analysis are considered.

The airflow inside school buildings office compartments with moderate environment, the evaluation of local thermal discomfort in a classroom equipped with crossed ventilation, the evaluation of comfort level in desks equipped with two personalized ventilation systems, the evaluation of thermal comfort conditions in a localized radiant system placed in front and behind two students seated nearby warmed curtains, the evaluation of thermal comfort conditions in a classroom equipped with cooling radiant systems and subjected to uniform convective environment, the application of a developed adaptive model in the evaluation of thermal comfort in ventilated kindergarten occupied spaces, the evaluation of indoor air quality in a classroom equipped with crossed ventilation and the comfort and airflow evaluation in spaces equipped with mixing ventilation and cold radiant floor are some examples of analysed and developed work in this chapter.

In: Natural Ventilation
Editor: Oscar T. Haynes
ISBN: 978-1-63482-414-9

Chapter 1

NATURAL VENTILATION IN BUILDINGS: AN OVERVIEW

Omar S. Asfour
Department of Architecture, Islamic University of Gaza, Palestine

ABSTRACT

Natural ventilation is a sustainable design strategy. However, its utilisation in buildings requires understanding of many relevant aspects including its driving forces and strategies. Natural ventilation is of a great environmental advantage, compared to the mechanical ventilation strategy. This has encouraged several researches on natural ventilation implementation in buildings, considering its main objective of providing buildings with the required air quality and quantity.

A review of human thermal comfort revealed that natural ventilation has a high potential for improving thermal comfort in hot climates, implementing wind-induced ventilation, and in cold climates, implementing temperature-induced ventilation. Within this context, natural ventilation can be used in several strategies and techniques for passive heating or cooling. Natural ventilation affects different heat transfer mechanisms between human body and its ambient environment. This mostly occurs in the form of sensible heat, by convection and radiation, and the form of latent heat, by evaporation of moisture on the skin.

INTRODUCTION

Since the industrial revolution, dramatic changes have occurred in the world of architectural design. The architect has stepped away from simple vernacular designs towards a design, which is characterised by being heavily energy consuming, in terms of construction and operation. For example, it is estimated that 50% of all resources consumed across our planet are used in construction (Edwards and Hyett, 2001). Large amount of this energy consumption is spent on the operation and control of buildings indoor environment, which is heavily dependent on mechanical systems in many types of buildings and in different ranges. This has resulted in an extensive consumption of unsustainable energy resources such as fossil fuels, which affect the quality of our built environment.

As the population of the world increases, dependence on these unsustainable energy resources is no longer sufficient nor reliable. Researchers today, and for many decades, have faced the challenge of introducing solutions that increase buildings dependency on natural and sustainable resources of energy. One of these challenges is to secure thermal comfort within the indoor environment of buildings relying on passive means and systems. Amongst these passive systems is ventilation provision in terms of air quantity and quality. It is an old problem, yet addressed by many innovative solutions. One of these is the straightforward and easy-to-use application of natural ventilation strategy.

This chapter aims to introduce a summarised presentation of natural ventilation applications in buildings. It defines the concept of natural ventilation in buildings, and compares natural and mechanical ventilation strategies. Then it presents some aspects of natural ventilation utilisation in buildings, which includes its objectives, driving forces, strategies, and role for passive cooling. It then demonstrates the relationship between natural ventilation and human thermal comfort. This includes the concept of human thermal comfort, its personal and environmental factors, in addition to an overview on the related heat transfer mechanisms. Then, it highlights the role of natural ventilation in improving thermal comfort conditions. This chapter concludes by comparing the advantages and disadvantages of natural ventilation use in buildings.

1. CONCEPT OF NATURAL VENTILATION IN BUILDINGS

Ventilation in any building can be one of two kinds (Liddament, 1996):

1.1. Purpose-Provided Ventilation

This is defined as "the process by which clean air, normally outdoor air, is intentionally provided to a space and stale air is removed" (Liddament, 1996, p.1). This process can be accomplished by a natural, mechanical, or hybrid system. Natural ventilation is defined as "the air flow resulting from the designed provision of specified apertures such as openable windows, ventilators, shafts, etc. and can usually be controlled to some extent by the occupant" (CIBSE, 1988, p.A4.1).

1.2. Infiltration

Infiltration term is used to describe the uncontrolled air movement to and from buildings. This air movement occurs naturally through cracks, gaps, and porous materials used in building facades (Figure 1).

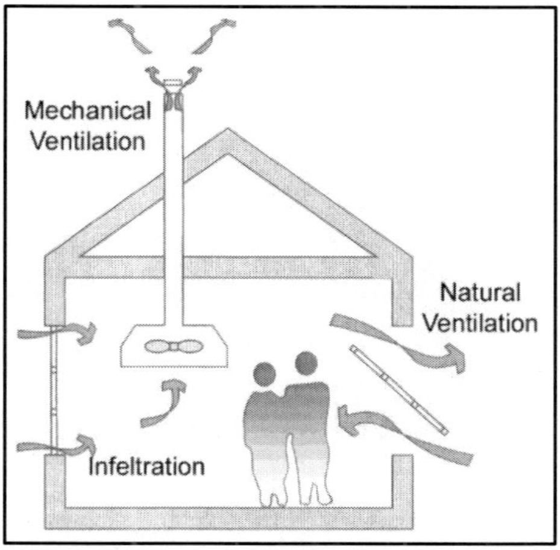

Source: (BICEPS module, 1997).

Figure 1. Ventilation is a combination of infiltration, and purpose-provided ventilation by natural or mechanical means.

2. Mechanical Ventilation

Although this chapter focuses on natural ventilation, it is useful to have a summarised idea about mechanical ventilation. This helps rationalising the advantages and disadvantages of each strategy, as will be discussed in Section 3. Mechanical ventilation relies on a mechanically generated power to drive airflow, instead of the natural driving forces. According to CIBSE (1997), mechanical ventilation strategy is divided into:

2.1. Mechanical Extract

In this system, a fan is used to extract air out of the space. Accordingly, the generated suction forces drive the outdoor air into the space through the purpose-provided openings.

2.2. Mechanical Supply

In this system, a fan is used to supply air into the space, where pressure difference forces the air to leave the space through some purpose-provided openings.

2.3. Mechanical Balanced Ventilation

Here, both supply and extract networks are incorporated into the space to provide filtered air and extract the 'old' one. It is possible that a proportion of the indoor air is re-circulated and mixed with the incoming outdoor air.

3. Why Natural Ventilation?

In the early stage of any design process, it is crucial to evaluate the feasibility of the use of natural ventilation strategy. The most important criterion in this regard is to deliver the required amount of air in the required time and place, with a minimal environmental impact. According to CIBSE

(1997), evaluation of advantages and disadvantages of any HVAC system is governed by many factors including:

- Robustness: the possibility of any failure should be considered.
- System cost: this includes: capital, operation, and maintenance costs.
- User's preference: this includes thermal comfort, quality of the indoor environment, and system control.

3.1. Natural Ventilation Advantages

Natural ventilation utilisation in buildings, compared to the mechanical ventilation, has many advantages. This includes (Energy efficiency Best Practice Programme, 1998):

- It is fossil fuel free. This has no direct negative environmental impact like air pollution, global warming, etc.
- It requires less construction and operation cost, and minimum maintenance.
- It is reliable and easy-to-use in many types of buildings such as residential and office buildings. The potential for personal control of the environment increases users satisfaction and productivity.

3.2. Natural Ventilation Use Disadvantages

Despite of the above-mentioned advantages of natural ventilation, it still has some disadvantages including:

- Wind has a random nature in both direction and magnitude. This means that securing stable indoor environment depending totally on natural ventilation can't be guaranteed, compared to the steady conditions of mechanical ventilation.
- Utilising natural ventilation in some building types seems to be impractical, especially those of deep plans, or those requiring high control level of indoor environment such as hospital buildings.
- Natural ventilation utilisation has many consequent issues like building security, safety, including the event of fire, and reliable control.

3.3. Balance between Advantages and Disadvantages

It is crucial in the early stage of the design process to balance between advantages and disadvantages of natural ventilation utilisation considering the conditions of each individual project. Thus, it is difficult to state an absolute rule here. However, and from an environmental point of view, the use of natural ventilation should be promoted wherever possible and feasible. As indicated in the introduction, mechanical ventilation has been heavily used in buildings in the last few decades.

This has caused a rapid increase in energy consumption in these buildings, which has to be rationalised and reduced. As a general rule of thumb, naturally ventilated buildings are 40% less in operation energy cost when compared to mechanically ventilated buildings (CIBSE, 1997). As a result, the concept of 'energy-efficient design' has developed the architectural design priorities. This design strategy aims to "provide thermal comfort and acceptable indoor air quality with the minimum use of energy" (CIBSE, 1998, p.6-1). Several studies have been carried out to make the use of natural ventilation in buildings more efficient, and reduce the use of mechanical plants (Liddament, 1996). This includes the following aspects:

- Producing natural ventilation design guidelines for different types of buildings.
- Improving the detailed design of architectural elements that can be utilised for natural ventilation.
- Utilisation of hybrid ventilation systems, wherever the use of natural ventilation as a sole strategy is impractical. These systems integrate natural ventilation with mechanical energy-efficient plants.

4. Objectives of Natural Ventilation in Buildings

Ventilation of buildings is important for health and comfort requirements at all building types. Natural ventilation has two main objectives: to control air quality, and to provide passive cooling or heating.

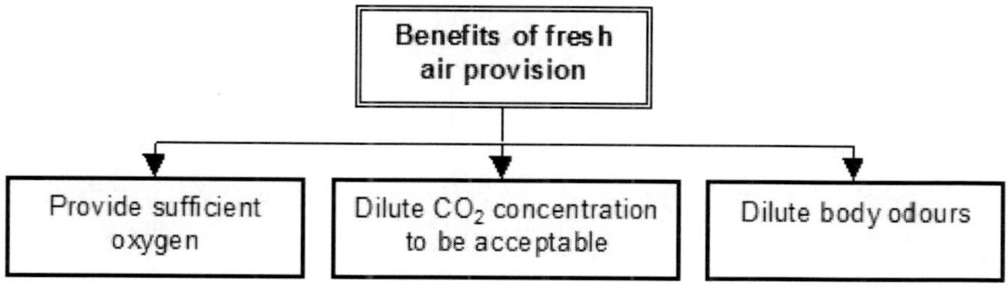

Source: (BRE, 1994). Reproduced by the author

Figure 2. Benefits of fresh air provision.

4.1. Natural Ventilation for Indoor Air Quality

Indoor Air Quality (IAQ) is an important side of natural ventilation application in buildings. In general, natural ventilation improves IAQ through three mechanisms, as illustrated in Figure 2.

It is important to note that little ventilation may result in a poor IAQ, while excessive ventilation may cause undesired draughts. Ultimately, ventilation rate should be sufficient to maintain both IAQ and thermal comfort requirements. In summer, sufficient oxygen is normally provided through windows. In winter, according to BRE (1994), background ventilation is required in a minimum rate of 400 mm^2/ m^2 of multi-cell building floor area. This can be achieved by using trickle ventilation. In addition, natural ventilation protects the environment from the harmful gases. One of the most common gaseous pollutants is Carbon Dioxide (CO_2).

Table 1. Limits of some internal gaseous pollutants in buildings

Pollutants	Source	Levels (ppm)	Limits (ppm)
Carbon Dioxide	People, combustion	320 to 2500	5000 ppm
Oxides of Nitrogen	Combustion, cooking, smoking	0.5 to 0.3	---
Carbon Monoxide	Combustion, cooking.	3 to 17	35 ppm

Source: (BICEPS module, 1997)

The effect of CO_2 concentration on human being varies from deep respiration to immediate death. Thus, it is important to predict the expected CO_2 concentration due to building occupants, equipment, external sources, etc., and insure that the upper allowed limit is not exceeded. The amount of CO_2 in air is usually expressed in percent, or in parts per million (ppm). Limits of some internal gaseous pollutants in buildings are listed in Table 1. Many solutions have been introduced to control this concentration like CO_2 concentration sensors. When CO_2 concentration reaches a predefined level, these sensors automatically operate natural ventilation by opening windows or roof vents to replace the stale air with fresh one.

4.2. Natural Ventilation for Passive Cooling

Passive cooling is defined as "processes of heat dissipation that will occur naturally, that is without the mediation of mechanical components or energy inputs" (Goulding et al., 1992, p.91). Many passive cooling techniques are available. Their implementation depends on the local conditions and design limitations. One of these techniques is natural ventilation. Figure 3 categorises and summarises some of these techniques.

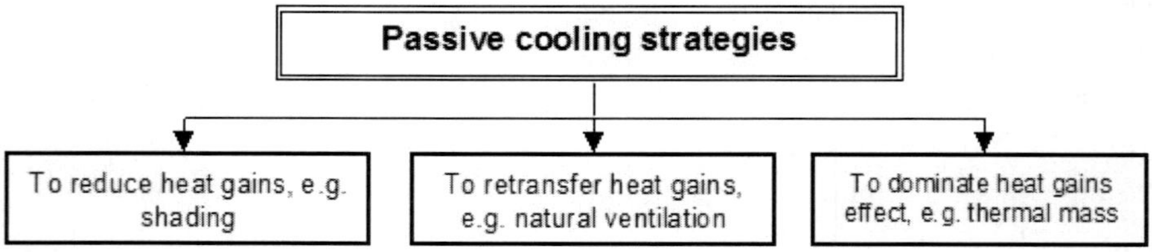

Figure 3. Natural ventilation is one of the passive cooling strategies.

Natural ventilation, as a passive cooling strategy, is used to cool down indoor air, building structure, and building occupants. The role of natural ventilation in passive cooling is highly affected by the local climatic conditions. For example:

- In summer: if the outdoor air temperature is less than the indoor one, airflow through a building removes internal heat gains and increase human thermal comfort. If the outdoor air temperature is higher than the indoor one, it is possible to reduce air temperature by some techniques such as night-time ventilation, explained in Section 6, and underground cooling. In hot arid regions, natural ventilation works effectively when integrated with an evaporative cooling method.

- If both outdoor and indoor air temperatures are equal, natural ventilation still important to supply fresh air and increase occupants comfort.
- In winter, a minimum ventilation, or infiltration, rate is required to maintain an acceptable indoor air quality.

5. Natural Ventilation Driving Forces

Natural ventilation relies on natural driving forces. These forces are either the stack effect, due to the difference in air temperature, or the wind effect, due to the difference in air pressure. These two driving forces are discussed below:

5.1. Wind Effect

In wind-induced or wind-driven ventilation, air passes through a building due to pressure difference generated between its inlets and outlets. When wind reaches the windward face of a building, a high-pressure zone formulates there. This pressure pushes air inside, around, and over this building. Accordingly, a negative pressure zone occurs on the literal and leeward sides (Figure 4).

This results in air movement from the higher-pressure zone to the lower one, which is known as the wind effect. Wind-induced ventilation is more effective in hot climates, when compared with stack-induced one. This is because of the low temperature difference between indoor and outdoor temperatures (Chow, 2004). Wind-driven ventilation depends on many factors.

5.2. Stack Effect

This kind of ventilation is known as the temperature-induced or buoyancy-induced ventilation. It occurs as a result of the difference between indoor and outdoor air temperatures. Warmer air has less density. Thus, and due to density stratification, this air rises to leave the building through the high-level openings, and the cooler outdoor air replaces it through building inlets or by infiltration (Figure 5). This is dependent on the stack height and temperature difference.

5.3. Combined Stack and Wind Effect

In some climatic conditions, both wind and stack driving forces have an effective role in ventilation. So, combining them is a useful strategy for thermal comfort improvement. It is important in this case to avoid any possible adverse effect of this combination, e.g. when they drive airflow in opposite directions.

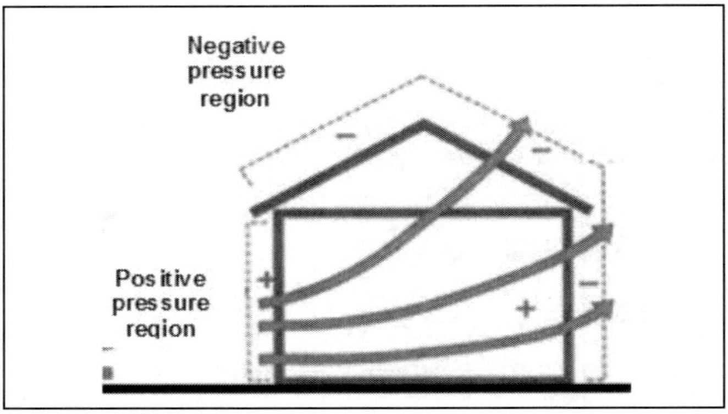

Source: (Liddament, 1996).

Figure 4. Wind-induced natural ventilation.

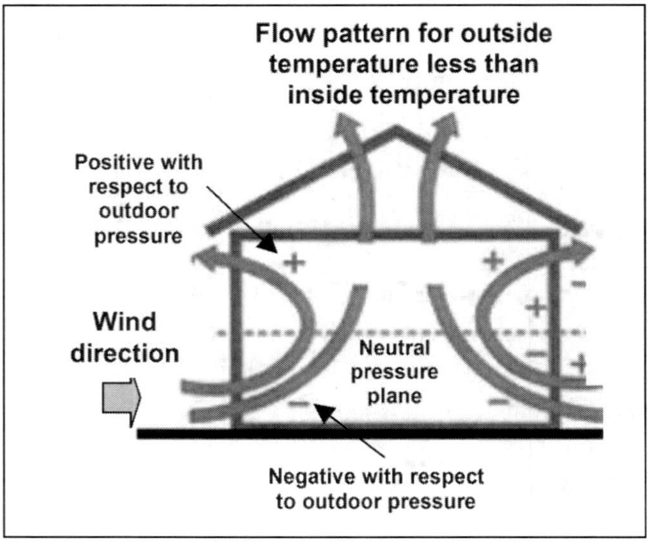

Source: (Liddament, 1996).

Figure 5. Temperature-induced natural ventilation.

6. Natural Ventilation Strategies

It is possible to implement natural ventilation in buildings in different systems. Each system has its own advantages and limitations of applicability, due to the local climatic conditions and design requirements. This section presents a summary about some of these systems.

6.1. Wind-Induced Ventilation

This strategy depends mainly on pressure difference between inlets and outlets as a driving force. It includes the following two strategies illustrated in Figure 6:

A. Single-Sided Ventilation

Single-sided ventilation, as illustrated in Figure 6, operates by openings located in one side of the building. As a general rule of thumb, it is effective up to a maximum depth equals to the double of a space height (CIBSE, 1997).

B. Cross Ventilation

Cross ventilation is more effective than single-sided ventilation strategy. It is used to create airflow paths between building inlets and outlets, which passes through the occupied level of the building. Cross ventilation is effective up to a maximum depth equals to five times the height of the room (CIBSE, 1997). It can be driven by wind effect, as in the case of using wall openings, or by stack effect, as in the case of using atrium ventilation, where the resulting flow path is vertical.

It is also important to plan good distribution of openings. Openings located at opposite walls are more effective for cross ventilation, compared to those located at adjacent walls. Also, vertical position of the inlet should be in relation with the occupied level of the space. Generally, low inlet is recommended for cooling and high outlet is recommended for stack ventilation.

The minimum window area should be about 5% of floor area, with the use of tickle ventilation to secure the minimum flow rate (Smith, 2001). In any case, window area should be enough to meet the required ventilation rate, which varies depending on the building use. For example, and according to CIBSE (1988, p.A1-9), the recommended airflow rate for open-plan offices is

8 l/s per person. Table 2 illustrates the three possible probabilities of inlet and outlet areas.

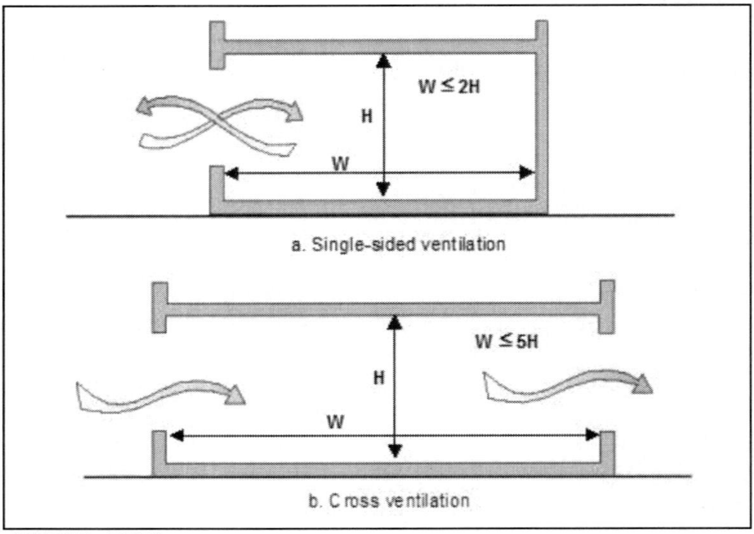

Source: (CIBSE, 1997).

Figure 6. Single-sided and cross ventilation strategies.

Table 2. Possibilities of inlet and outlet opening areas

Illustration (A_i =inlet area, A_o=outlet area)	Observation	Design target
$A_i = A_o$	Maximum air change created	Building cooling
$A_i < A_o$	Maximum interior air speed	Users cooling
$A_i > A_o$	Air speed reduced inside, and accelerated outside.	Outdoor spaces cooling.

Source: (Moore, 1993)

6.2. Solar-Induced Ventilation

This strategy depends mainly on temperature difference between inlets and outlets as a driving force. It includes:

A. Solar Chimney

The main components of the solar chimney are illustrated below. Solar chimney can be integrated in vertical elements of buildings such as stairwells and elevator shafts, chimneys, etc. Its design includes a black metal absorber installed behind its front. This is insulated from the internal space by the means of air cavity and thermal insulation. Heated air in this cavity rises and leaves through the top opening. Thus, cooler fresh air, provided by wall openings, is induced inside the space. This air circulation is accelerated when prevailing wind currents passes over the chimney. Solar chimney can also be utilised for heating in winter. This can be achieved by modifying openings arrangement, so that the heated air in the cavity is admitted inside the space, as illustrated in Figure 7.

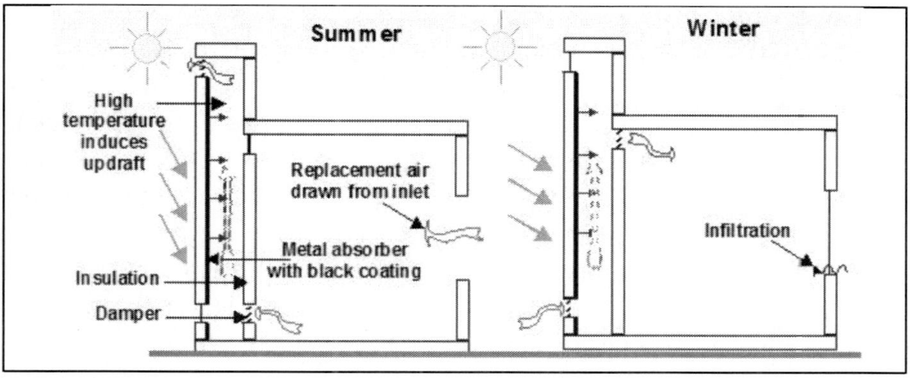

Figure 7. Solar chimney.

B. Trombe Wall

The main components of the Trombe wall are illustrated in Figure 8. Its operation concept is similar to that of solar chimney, but using different building materials with different thermal properties. Trombe wall can be utilised effectively for heating in winter by utilising solar radiation during daytime. During night, and depending on the thermal properties of the wall located beyond the glazing panels, heat stored in the wall is released, which heats up the space.

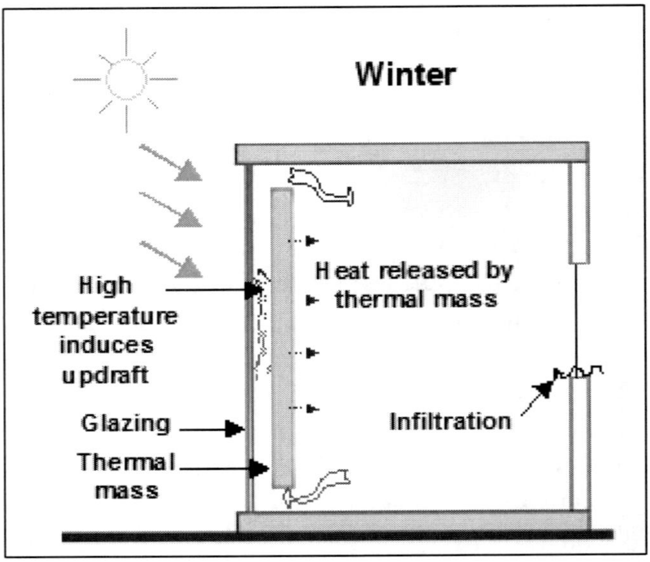

Figure 8. Trombe wall.

C. Atrium Ventilation

Atrium building is simply is a glass-covered courtyard, towards which all internal spaces are oriented. This architectural configuration is widely used in public buildings like offices. Airflow is induced from the different building spaces towards the atrium by stack effect. Air then rises to the top of the atrium, where outlets are placed. In this case, the atrium should be extended above the roof by several meters to increase stack height for the upper floor. The difference between atrium and solar chimney is that atrium function as a central space for circulation and social gathering. This makes it an attractive ventilation strategy in public buildings.

D. Double Façades

Double facades are used to induce air through a cavity located between glass panels in the outer skin of the building. This is implemented for the whole façade instead of a separated vertical element as in Tromb wall and the solar chimney. Air in the cavity is heated up and driven inside the building by stack effect, with or without the use of fans. In summer, hot air leaves the cavity through the outer layer openings. The application of other ventilation strategies including night ventilation is possible, depending on the type and orientation of the double skin façade.

6.3. Night Ventilation

This strategy relies mainly on the thermal properties of building materials, which are used as a thermal sink. The low temperature of night airflow cools down building envelope, when this airflow passes over it. At morning, this envelope absorbs heat gains, released due to solar radiation and building operation. Thus, efficiency of night-time ventilation, as a passive cooling technique, depends on two factors:

- Difference between the outdoor and indoor air temperature.
- Thermal characteristics of the building materials used.

According to CIBSE (1997), some issues should be considered when applying this strategy including:

- Appropriate building material selection, with sufficient contact surface area between them and air.
- Security of the building.
- Temperature control to avoid over-cooling.

6.4. Mixed Natural Ventilation Strategy

Different natural ventilation systems can be combined in one building. This gives the designer a wider margin to deal with each single space according to its requirements. As an example, simple combination between cross-ventilation, for an open plan office area, and single-sided ventilation, for cellular office rooms, is possible.

6.5. Mechanically-Assisted Strategy

In some cases, natural ventilation becomes insufficient to secure an acceptable level of human thermal comfort. It is possible to overcome this problem by integrating some mechanical devices such as fans to assist the use of natural ventilation, as illustrated in Figure 9. In mechanically ventilated buildings, this strategy is useful to reduce energy consumption. This requires an assessment of the local climate conditions, in addition to an effective control strategy. Some examples of the application of this strategy are:

- The use of fans at the top of solar chimney to help air extract, when driving stack is not enough to achieve the required air change rate.
- The use of mechanical ventilation for some spaces located in the centre of deep-plan buildings.
- The use of mechanical supply when quality of the outdoor air is unacceptable.

6.6. Criteria of Natural Ventilation Strategy Selection

Depending on local climatic conditions and building design requirements, it is possible to choose an appropriate natural ventilation strategy. These factors are illustrated in Figure 10:

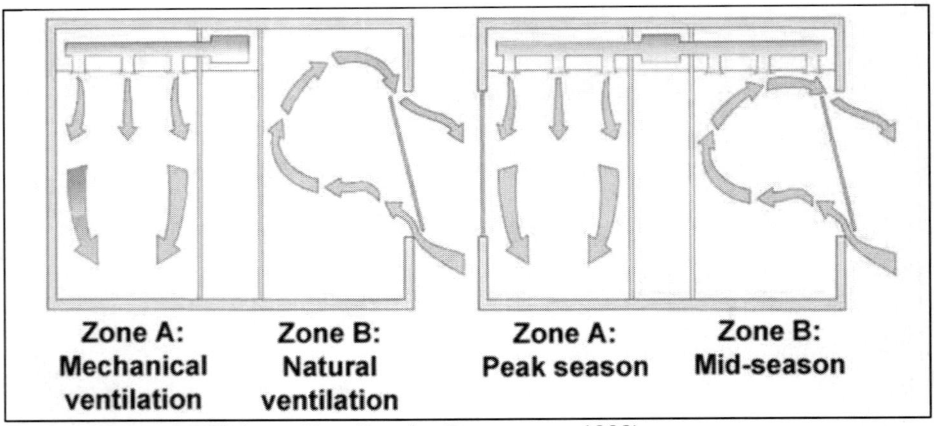

Source: (Energy Efficiency Best Practice Programme, 1998).

Figure 9. Examples of hybrid ventilation.

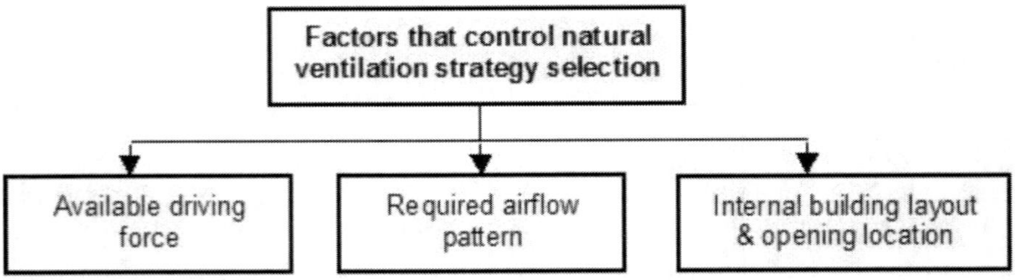

Figure 10. Factors that control selection of natural ventilation strategy.

As for the first factor, local climatic data analysis helps specifying the dominant ventilation driving force. In the second factor, assessment of the resulting airflow pattern is mainly qualitative. It depends on the requirements of building operation and users. Some related aspects here are direction of airflow paths from the inlets towards the outlets, maximum airflow speed, airflow behaviour in the case of fire, etc. Finally, the third factor is related to the layout of the plan and the location of its openings.

7. Natural Ventilation and Thermal Comfort

This section presents an overview on human thermal comfort in buildings, and the role of natural ventilation in this regard.

Table 3. Typical heat emission of human body against its activity

Activity	W/m^2	*met*
Sleeping	40	0.7
Seated	60	1.0
Walking	150	2.6
Exercise	175-235	3.0-4.0

Source: (Goulding et al., 1992). Summarised by the author.

7.1. Human Thermal Comfort Definition

There are many definitions of thermal comfort. This is because thermal comfort is dependent on many parameters, which are sometime difficult to measure, and depend on personal preferences. For example, this is clear in ASHRAE Fundamentals (1997) (see: Santamouris and Asimakopoulos, 1996, p.129), which defines human thermal comfort as "the conditions in which a person would prefer neither warmer nor cooler surroundings". The same point is also clear in the definition given in the British Standards (BSI, 2005, p. 10), which states that thermal comfort is "that condition of mind which expresses satisfaction with the thermal environment".

Furthermore, thermal comfort control is more difficult in naturally ventilated buildings duo to the instability in the climatic conditions. Despite this difficulty, many models have been developed for thermal comfort assessment. To understand the assessment criteria, it is crucial to demonstrate the parameters that affect human thermal comfort, and the related heat transfer mechanisms.

7.2. Thermal Comfort Personal Factors

These factors are related to human beings such as age, sex, and activity. As indicated by Santamouris & Asimakopoulos (1996), the most important factors are:

A. Occupant Activity

Metabolic rate of human body is in relation with its activity. For example, a person doing a heavy activity requires more cooling, and vice versa. This parameter estimates the amount of energy produced per square metre of body surface area of human beings (Table 3). It is measured in metabolic or *met* unit, which is "the metabolic rate of a seated person when relaxing, i.e. 58 W/m^2" (Goulding et al., 1992, p.60).

B. Occupant Clothing

Clothing can be considered as a bodily thermal insulation. The unit of this insulation is *clo*, which is equivalent to an insulation of 0.155 m^2 K/W obtained by a winter business suit. Table 4 illustrates thermal insulation of different clothing categories. Similarly to the previous factor, clothing affect heat transfer rate between the body and the ambient environment.

7.3. Thermal Comfort Environmental Factors

There are many environmental factors that affect thermal comfort, as shown in Figure 11. A summary on each factor is given below.

Table 4. Clothing values for thermal comfort calculations

Clothing	m^2 K/W	*clo*
Nude	0	0
Light summer ensemble	0.08	0.5
Typical indoor winter ensemble	0.16	1.0
Heavy business suit	0.23	1.5

Source: (Goulding et al., 1992). Summarised by the author.

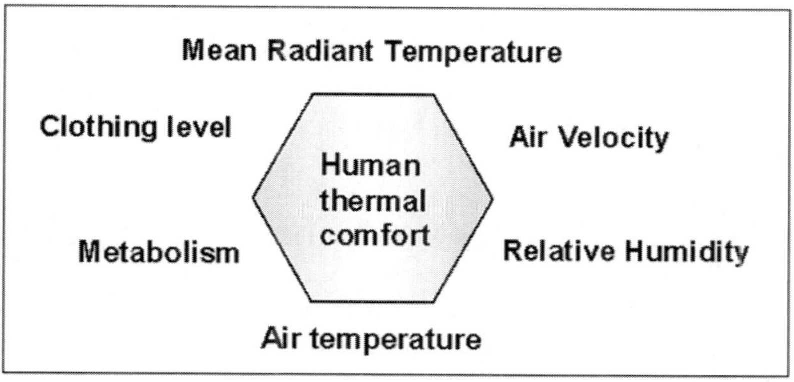

Figure 11. Primary factors affecting human thermal comfort.

A. Dry-Bulb Temperature, T_{db}

T_{db}, or air temperature as usually named, is the ambient air temperature measured by a thermometer freely exposed to heat radiation in the air. It is called so because it is measured with a standard thermometer whose bulb is not wet. This is to avoid any effect of evaporative heat transfer, which is considered in the Wet-Bulb Temperature, T_{wb}. It is usually measured in degrees Celsius (°C) or degrees Farenheight (°F).

B. Mean Radiant Temperature, MRT

Objects in the space affect human thermal comfort, even if they are not in a direct contact with the human body. This is considered by the environmental factor of MRT. These objects absorb or emit heat, depending on their thermal properties and temperature difference between them and the surrounding air. Thus, MRT refers to the average temperature of surrounding surfaces, which is nearly the same of ambient air temperature in the normal cases (Moore, 1993).

MRT is measured using globe thermometer, which is a normal dry bulb thermometer encased in a 150 mm diameter matt-black copper sphere which has absorptivity approaching that of the skin. When surrounding objects are warmer than the average skin temperature, MRT should be positive, and vice versa. It is important to note that MRT is independent of ambient air temperature. For example, if the body is exposed to the sun it will be more comfortable even at lower air temperatures.

C. Relative Humidity, RH

Humidity refers to the amount of moisture vapour in a specific volume of warm air. This amount of vapour, under any dry bulb temperature, is known as the absolute humidity, and is measured in g/kg. Maximum possible amount is known as dew point. Relative humidity (RH), therefore, is the ratio between the absolute humidity to this maximum amount, expressed as a percentage. Relative humidity is measured using hygrometers.

D. Air Velocity

Air movement can be natural or mechanical. It is measured using anemometers, normally in metres per second. Wind velocity is defined as "a constantly varying vector quantity described by its speed and direction" (Goulding et al., 1992, p.32). If its speed is expressed in m/s, air pressure in equation 1 will be expressed in kg/m/s^2, i.e. N/m^2 or Pascal (Pa). Wind speed is recorded and averaged in meteorological stations for a defined period of time. Wind velocity at any point in this record can be estimated from (Awbi, 2003):

$$V_{(t)} = V + v_{(t)} \quad (1)$$

where $V_{(t)}$ is the instantaneous wind speed, V is the time-mean value, and $v_{(t)}$ is the fluctuating component. By integrating $V_{(t)}$ over a known period of time, V can be estimated.

7.4. Bodily Heat Transfer Mechanisms

Any building can be described as a "set of various systems coupled together" (Allard & Alvarez, 1998, p.30). Amongst these systems are the heat transfer mechanisms. According to Marsh (1998), physiologists found that absorptivity and emissivity of human skin is considered high, compared to other known substances. This means that human body is highly responsive to any changes in the ambient temperature.

To understand the effect of natural ventilation, as a passive cooling technique, on human thermal comfort, it is essential to explain how human body gives off heat by the skin surface and respiration. This is because any passive cooling technique that utilises natural ventilation has in its heart the implementation of heat transfer mechanisms. Bodily heat transfer mechanisms can be divided into four mechanisms (Moore, 1993):

A. Convection

Heat energy here is transferred through fluids by the physical movement of particles, e.g. movement caused by buoyancy effect. Convective heat transfer rate is in proportion with temperature difference between skin and ambient air. It is also in proportion with air speed. This means that skin loses heat by convection when cold air passes over it. Also, additional convection occurs due to breathing as cool air is drawn into the body.

B. Radiation

Heat energy here is transferred by electromagnetic waves. This occurs when surface atoms of any material emit thermal energy as electromagnetic waves in the infrared range of wavelength. Radiation rate is dependent on material properties and temperature. Human skin loses heat by radiation depending on the MRT value of surrounding objects. This is also dependent on the distance between these objects and the human body.

C. Evaporation

When skin temperature increases, due to the increase of ambient temperature or activity level, it becomes difficult for the body to maintain its heat balance depending only on convection and radiation. In this case, evaporation of sweat encourages more heat transfer between skin and ambient air, causing skin to cool down. This mechanism is in inverse proportion to relative humidity of the ambient air. This means that in 100% RH, no moisture absorption occurs in the air, and hence no evaporation. Evaporation also is in proportion to the ambient air velocity.

D. Conduction

Heat energy here is transferred through materials without molecules movement. This mechanism occurs in both solid materials and fluids. Heat here is transferred directly between adjacent molecules in a time duration, which depends on material properties. This mechanism is less important than the above-mentioned ones in terms of thermal comfort determination It occurs when human body is in a direct contact with another material, which has effective thermal capacity and conductivity, like ground tiles for instance.

7.5. Effect of Natural Ventilation on Thermal Comfort Environmental Factors

As mentioned before, one of the main objectives of natural ventilation is to encourage heat transfer between human body and its ambient environment. The effect of natural ventilation on the environmental factors related to human thermal comfort is highlighted below:

A. Effect of Air Velocity on Air Temperature

It is recommended to reduce air velocity in winter in order to maintain the required air temperature for thermal comfort. In summer, and when outdoor air temperature is less than the indoor temperature, increasing air velocity by natural ventilation encourages bodily heat loss by convection, which reduces the ambient air temperature. This increases convective heat transfer between skin and ambient air, as cooler air replaces the warmer one, which is closer to the skin.

If the outdoor temperature is already higher than upper recommended limit for thermal comfort, then the use of thermal mass and night ventilation, as explained in Section 6, are useful strategies to improve thermal comfort conditions by both convection and radiation. The above-mentioned two scenarios are true in the case of hot arid climate. In the case of hot humid climate, high air velocities help enhancing evaporative heat transfer over skin surface. This is because drier ambient air replaces the saturated air near the skin. In all cases, air velocity should be kept below its allowed upper limit. Table 5 gives some indications in this regard:

B. Effect of Air Velocity on MRT

MRT is a very important factor that determines human thermal comfort. In winter, increasing MRT can compensate the low value of air temperature. In summer, and for passive cooling, increasing air velocity helps removing the undesired heat emitted by the ambient surfaces. Thus, these surfaces are cooled down by convection. When MRT becomes less than skin temperature, these surfaces act as thermal sinks. This can be enhanced by implementing night-time ventilation, provided that outdoor temperature is low enough.

Table 5. Internal air speed classification in buildings

Airflow description	Airflow velocity		Airflow description	Airflow velocity	
	m/s	km/h		m/s	km/h
Still	0.0	0.0	Hair and Papers Move	1.0	4.0
Not Noticeable	0.1	0.4	Noticeably Draughty	1.4	5.0
Barely Noticeable	0.3	1.0	Unpleasant Breeze	1.7	6.0
Pleasant Breeze	0.5	1.8	Gusting Breeze	2.0+	6.5+
Light Breeze	0.7	2.5	---	---	---

Source: (Marsh, 1998)

C. Effect of Air Velocity on RH

Relative humidity is usually recommended to be between 40% and 70%. Values outside this range result in the climate being humid or arid, in addition to being hot. According to ASHRAE, (see: Goulding et al., 1992), when air temperature is more that 30°C, the effect of RH on thermal comfort becomes more pronounced. In high relative humidity, increasing air movement effectively encourages bodily heat transfer by evaporation.

7.6. Thermal Comfort Assessment

Thermal comfort prediction is not an easy task due to the variety of interacting factors, including personal preferences. However, many models have been developed. In fact, explaining these models is out of the scope of this chapter, as the main objective here is to give a summarised background. As explained by Awbi (2003), the well-known Fanger's model, developed in 1970, is the most comprehensive model for thermal comfort assessment up to date. Fanger's steady-state energy balance model applies the heat balance equation, in addition to other empirical equations and different assumptions. This equation is:

$$S = M + W + R + C + K - E - RES \quad (2)$$

where: S = heat storage in the body, M = metabolic rate, W = mechanical work, R = heat exchange by radiation, C = heat exchange by convection, K = heat exchange by conduction, E = heat loss by evaporation, and RES = heat loss by respiration (all units are in W).

Table 6. Fanger's PMV index

Thermal description	PMV
Hot / cold	±3
Warm / cool	±2
Slightly warm / cool	±1
Neutral	0

Source: (Santamouris & Asimakopoulos, 1996)

This equation is derived depending on the Law of Energy Conservation. It balances the rate of bodily metabolic and activity heat production to the heat loss by different mechanisms. To describe thermal sensation when human body is not in thermal equilibrium, Fanger's model estimates what is known as the Predicted Mean Vote. PMV index contains many categories to describe mean thermal sensation of a large group of people. This scale is recommended by ISO to be between –0.5 and 0.5.

It is also possible to use PMV calculators to investigate the effect of the different thermal comfort factors. Knowing the PMV, the Predicted Percentage of Dissatisfied people (PPD) can be determined as well. PPD increases as PMV moves up or down from 0, which is the neutral value. The maximum value of PPD is 100% and the minimum value is 5%, as it is impossible practically to satisfy all building users.

References

Allard, F. & Alvarez, S. (1998). Fundamentals of natural ventilation. *In*: Allard, F., ed. *Natural ventilation in buildings: a design handbook.* London: James & James Ltd., 1998, p. 9-62.

Awbi, H. (2003). *Ventilation of buildings.* 2nd ed. London: Spon Press.

BICEPS module. (1997). *Ventilation: energy efficiency and indoor air quality illustrations*'. Watford: Building Research Establishment.

BRE. (1994). *Natural ventilation in non-domestic buildings.* BRE Digest 399. Garston: Building Research Establishment.

BSI. (2005). BS EN ISO 7730: *Ergonomics of the thermal environment – Analytical determination and interpretation of thermal comfort using calculation of the PMV and PPD indices and local thermal comfort criteria*. London: British Standards Institution.

Chow, W. K. (2004). Wind-induced indoor-air flow in a high-rise building adjacent to a vertical wall. *Applied Energy*, *77*, 225-234.

CIBSE. (1988). *CIBSE Guide*. Vol. A. and B. 5th ed. London: the Chartered Institution of Building Services Engineers.

CIBSE. (1997). *Natural ventilation in non-domestic buildings*. Application manual AM10. London: the Chartered Institution of Building Services Engineers.

CIBSE. (1998). *Energy efficiency in buildings*. London: the Chartered Institution of Building Services Engineers.

Edwards, B. & Hyett, P. (2001). *Rough guide to sustainability*, London: RIBA Publications.

Energy Efficiency Best Practice Programme. (1998). *Natural ventilation in non-domestic buildings - a guide for designers, developers and owners*. Good practice guide 237. London: the Chartered Institution of Building Services Engineers.

Goulding, J. R., Lewis, J. O. & Steemers, T. C. (1992). *Energy in architecture: the European passive solar handbook*. London: Batsford Limited.

Liddament, M. W. (1996). *A guide to energy efficient ventilation*. Coventry: Air Infiltration and Ventilation Centre.

Marsh, A. (1998). *Efficient design: human comfort* [online]. Available at: <http://www.squ1.com/comfort/> [Accessed 15 September 2004].

Moore, F. (1993). *Environmental control systems: heating cooling lighting*. New York: McGraw-Hill, Inc.

Santamouris, M. (1998). Prediction methods. *In*: Allard, F., ed *Natural ventilation in buildings: a design handbook*. London: James & James Ltd. p. 63-158.

Santamouris, M. & Asimakopoulos, D. eds, (1996). *Passive cooling of buildings*. London: James & James Ltd.

Smith, P. F. (2001). *Architecture in a climate of change*. Oxford: Architectural Press.

In: Natural Ventilation
Editor: Oscar T. Haynes
ISBN: 978-1-63482-414-9

Chapter 2

Thermal Comfort and Energy Efficiency in a Naturally Ventilated Office: CFD, BTM and a Field Study

John Kaiser Calautit[1,*], Sally Salome Shahzad[2†], Ben Richard Hughes[1‡], John Brennan[3,§] and Dimitris Theodossopoulos[3#]

[1]Department of Mechanical Engineering,
University of Sheffield, Sheffield, UK
[2]College of Engineering and Technology, University of Derby, Derby, UK
[3]Edinburgh School of Architecture and Landscape Architecture,
University of Edinburgh, UK

Abstract

The energy crisis drives the design of the workplace towards passive systems, such as natural ventilation. The design of the ventilation system influences the energy demand of the building and comfort of the user. In order to improve the design, optimise the energy performance, predict and improve users' comfort, the application of accurate computational

---

* J.K.Calautit@gmail.com.
† s.shahzad@derby.ac.uk.
‡ Ben.Hughes@sheffield.ac.uk.
§ John.Brennan@ed.ac.uk.
# D.Theodossopoulos@ed.ac.uk.

modelling and analysis techniques are essential. This study reviews the advances in thermal comfort modelling and energy analysis of the workplace using Computational Fluid Dynamics (CFD) and Building Thermal Modelling (BTM). This is followed by a case study of CFD-BTM analysis of energy and thermal comfort of a practice example of a naturally ventilated office. This is a three storey building with an open plan layout in the UK and the office area of 3000 m^2, including 375 workstations. It was awarded by the British Council for Offices, received an 'excellent' BREEAM rating and 10 out of 10 Environmental Performance Indicator (EPI) rating. The office is naturally ventilated with manually and mechanically operated windows and a stack effect through the vents above the atrium. Furthermore, environmental measurements and comfort surveys are applied in the case study buildings.

Keywords: Building Thermal Modelling (BTM), Computational Fluid Dynamics (CFD), open plan office, thermal comfort

1. Introduction and Case Study Description

This study investigated the accuracy of predicting user comfort through simulation. An actual building was analysed using Computational Fluid Dynamic (CFD) and Building Thermal Modelling (BTM). In order to validate the results, they were compared to the collected field data regarding the energy use of the building and occupants' view of thermal comfort.

Figure 1. East façade and floor plan.

Figure 2. West façade.

The selected building for this study was a British open plan office in Scotland, in which natural ventilation was in operation. It received an excellent BREEAM rating (84%), the highest Environmental Performance Indicator (EPI) rating (10 out of 10), and won an award from the British Council for Offices.

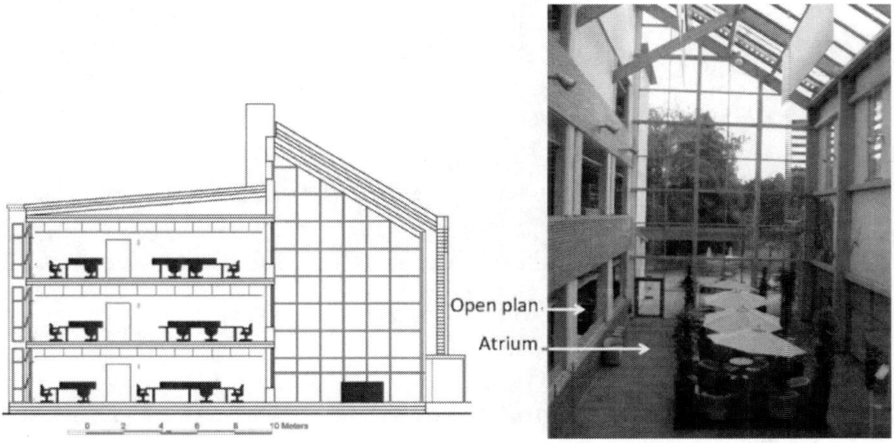

Figure 3. East-West Section and atrium.

Sustainable strategies were in place in terms of the use of recycled, low carbon and locally resourced materials, energy and thermal performance of the building as well as the overall design and natural ventilation. Biomass boiler and solar panels were used in the building as part of the energy strategy. The use of public transport and cycling were encouraged and provided for commuting to work. This was a three-storey office building with an open plan layout and 1000 m^2 and 125 workstations per floor area, where mainly administrative and sedentary activities took place. The location of the building and the design of a shallow open plan layout and a large glass façade allowed access to natural light, natural ventilation and an outside view of the city and the green fields for occupants, as presented in Figure 5.

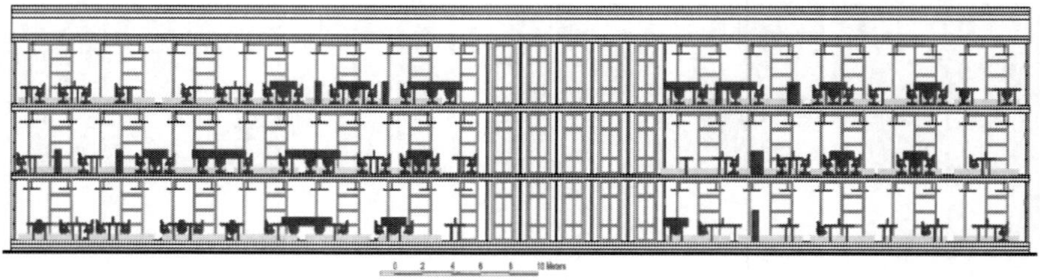

Figure 4. North-South Section.

Figure 5. View from the windows, window seat and far from the windows.

1.1. Ventilation and Thermal Systems

Natural ventilation was the main operating system and two types of windows were provided. The bottom windows were openable and provided for occupants seated around the perimeter of the eastern side of the building to control the thermal environment. The top windows presented in Figure 6 were mechanically controlled and connected to the temperature and carbon dioxide monitor devices in order to ensure a good quality of indoor air and temperature.

Figure 6. Mechanically controlled windows.

This centrally controlled system was also connected to several vents located above the atrium to increase the stack ventilation effect in the building, as illustrated in Figure 7.

Blinds were provided for occupants seated close to the atrium in order to control draughts and natural light from the skylight above the atrium, as illustrated in Figure 8.

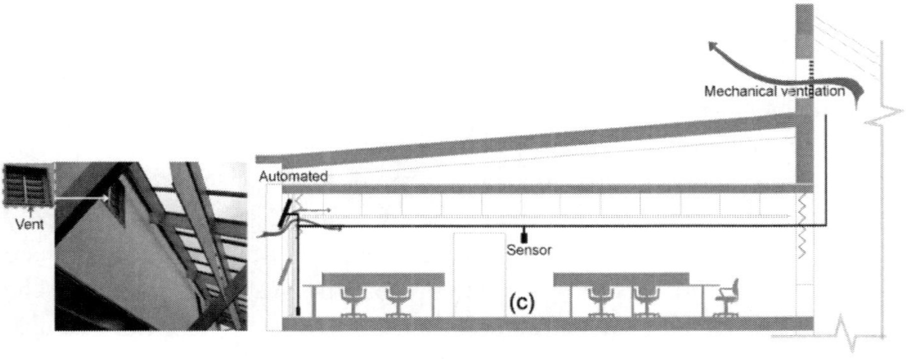

Figure 7. Vents above the atrium and section of centrally operated mechanical ventilation.

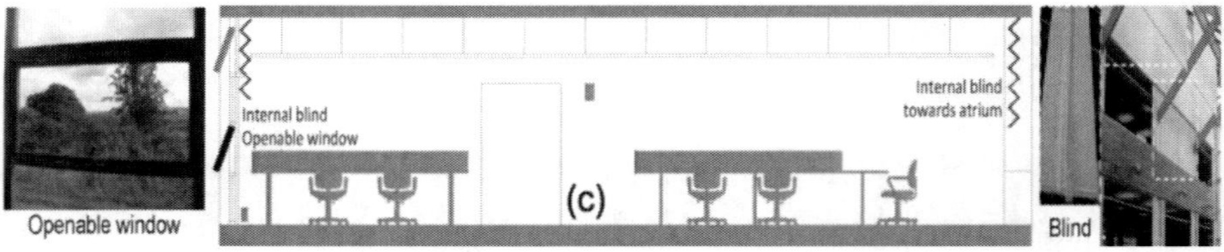

Figure 8. Blinds toward the atrium to prevent glare, solar gain and draughts.

Occupants seated around the eastern perimeter of the building were provided with openable windows and blinds to control the natural ventilation and light. Radiators around the eastern perimeter of the building reduce the radiation effect of the body loosing heat to the cold surface of the window.

2. Previous Related Work

2.1. Airflow and Comfort Analysis Using CFD

Computational Fluid Dynamics (CFD) modelling is a very useful method in evaluating and optimising the design of natural ventilation systems in buildings. Due to the sophisticated design of CFD tool, it enables a numerical model of a building to be studied under many different design constraints. CFD is commonly applied for the analysis of airflow process in buildings and thermal comfort studies in various climatic conditions. Many studies used CFD to predict useful variables related to indoor comfort, such as air velocity, pressure, temperature and relative humidity. The basis for CFD analyses of ventilation flows are the incompressible Navier–Stokes equations. These equations describe the motion of a viscous Newtonian fluid. Since there are no analytical solutions for these equations for the complex geometries typically these equations are solved numerically. The success of CFD has been partly due to the rapidly increasing capabilities of computing power and also progresses in numerical investigation and simulation (Etheridge, 2011). This also led to the formation of guidelines for attaining acceptable quality (Nielsen, 2004). The outcomes of the CFD simulation can be used to compare with the empirical results for validation purpose. In fact, it was found that the simulation outcomes often agreed well with the empirical results. Nevertheless, it should be noted that because of problems in defining the actual boundary conditions of the current available CFD software and possible adaptation and acclimatization of occupants to their thermal environment, some discrepancies with the empirical results may occur.

2.2. Comfort and Energy Analysis Using BTM

Many studies (Short et al., 2004, Jones-Lee and Loomes, 2003) have focused on the promotion of building energy efficiency. They revealed the significance of considering the built environment as a whole system to improve the design parameters simultaneously, since these parameters are strongly integrated. Several methods have been developed for predicting the energy consumption of buildings such as statistics-based tools and building thermal modelling (BTM). BTM is an extensive tool for designing and carrying out thermal simulations and analysis of rooms or entire buildings. BTM software perform under a simplified analytical method that demands input data such as properties of materials, geometric factors, climatic data, and heat transfer coefficients (Shi and Yang, 2013). An example of a commercial BTM software is the Autodesk Ecotect. Ecotect is a complete design and environmental analysis tool for buildings. It includes the full range of simulation and analysis functions required to assess the performance and operation of a building design (Schueter and Thessling, 2008). The analysis capabilities of the Ecotect software include energy and thermal analysis, and lighting analysis.

Yang et al. (2014) used the Ecotect analysis software for the planning of a residential state Ma'anshan City. The work analysed the climatic data and energy-saving methods that are related to local environment. Optimisation factors such as orientation of the model, daylighting, and natural ventilation were investigated. The authors concluded that the Ecotect software can assist architects and engineers in the early design stage of the building. The results provided by the software are useful for improving the energy efficiency of the building and creating a comfortable indoor environment. Shoubi et al. (2014) used the Ecotect software to evaluate several combinations of materials and identify alternate sustainable design solutions to reduce the energy consumption of a residential building. The authors concluded that simulating the performance of the building in the pre-construction stage can assist decision makers, engineers, and architects to select specifications with the least negative effect to the environment. Chen et al. (2014) investigated the daylighting performance in an industrial building. The simulation of the daylighting illuminance distributions was carried out using the Ecotect and

Desktop Radiance simulation programs. The results showed that the simulated daylighting illuminance matched well with the field measurements.

2.3. Field Studies of Thermal Comfort

Fidel (1984) states that 'field studies are investigations of phenomena as they occur without any significant intervention of the investigators' (Fidel, 1984). This is a systematic approach (Bromley, 1990) to investigate the phenomenon in its natural context using multiple sources of evidence and methods (Johansson, 2003).The aim of field studies of thermal comfort is to improve human comfort in buildings through developing comfort prediction models using the results of the field studies (Nicol and Roaf, 2005). Occupants' response to the thermal environment is subjective and according to Nicol (2004) 'the aim is often to predict the subjective impression from a knowledge of the physical environment' (Nicol, 2004). The context and time are important factors in field studies to capture occupants' instant responses to the thermal environment in their familiar everyday life situation with their normal clothing and behaviour (Nicol, 2004)(Nicol and Roaf, 2005).

Various methods have been applied in field studies of thermal comfort, the most common of which are environmental measurements and survey questionnaires. Bedford (1936) applied a survey questionnaire in several factories (Bedford, 1936) and Nicol and Humphreys (1973) applied field studies of thermal comfort at the natural context (Nicol and Humphreys, 1973b). The ASHRAE seven-point scale thermal sensation question (ASHRAE, 2009) is widely used in thermal comfort studies, presented in section 3.3 and a 'neutral thermal sensation' response is considered as a thermally comfortable status (ASHRAE, 2010). Operative temperature (combination of dry bulb temperature and mean radiant temperature) and relative humidity are considered the main thermal factors (ASHRAE, 2013). Air temperature is considered as the most important factor influencing human comfort (Auliciems and Szokolay, 1997). Nicol and Humphreys emphasised the importance of the relationship between indoor and outdoor conditions particularly in naturally ventilated buildings (Humphreys, 1977, Humphreys, 1978, Humphreys and Nicol, 1995, Nicol et al., 2012).

Computational Fluid Dynamics is used in thermal comfort studies to predict and improve the comfort conditions in buildings. Ctalina et al. applied a combination of CFD and experiments to evaluate predicted mean vote thermal comfort in an experimental chamber when radiant cooling was in operation (Catalina et al., 2009). Mainly environmental measurements were applied to calculate the PMV and mean radiant temperature in the room. Although the results of this study are useful and there are advantages in experimenting in a controlled environment, findings of experimental chambers often do not apply to the context of daily life (Nicol et al., 2012). Therefore in the present study, the results of CFD and BES were compared to the results of the real life context investigating occupants' views of the thermal environment through the application of the field studies of thermal comfort.

3. Methodology

3.1. Computational Fluid Dynamics (CFD)

Computational Fluid Dynamics (CFD) is a tool that uses numerical methods and algorithms to solve and analyse fluid flow problems. In building design, it can be used to predict air speed, pressure and temperature that will occur at any point throughout a predefined air volume in and around buildings with defined boundary conditions. CFD models are usually preferred for the assessment of air velocity since they can provide detailed information on the performance of mechanical or natural ventilation strategies. CFD has been implemented as the primary analysis tool in a number of studies concerning building ventilation performance.

In the present study, the CFD code from the ANSYS 12 Fluent software was employed. It used a primitive variable method involving the solution of a set of equations that described the conservation of heat, mass and momentum using the Navier–Stokes equations and the standard k-ε turbulence model. The Reynolds-Averaged Navier-Stokes (RANS) equations for mass, momentum, energy and species were solved using the commercial CFD code for the velocity and temperature field simulations. The model employed the control-volume technique and the Semi-Implicit Method for Pressure-Linked Equations (SIMPLE) velocity-pressure coupling algorithm with the second order upwind discretisation. The equation for conservation of mass, or continuity equation is as follows:

$$\frac{\partial \rho}{\partial t} + \nabla \cdot (\rho \vec{v}) = 0 \quad (1)$$

where *p* is the static pressure and is the velocity of the fluid. Transport of momentum in an inertial (non-accelerating) reference frame is described by:

$$\frac{\partial}{\partial t}(\rho\vec{v}) + \nabla\cdot(\rho\vec{v}\vec{v}) = -\nabla p + \nabla\cdot(\overline{\overline{\tau}}) + \rho\vec{g} + \vec{F} \tag{2}$$

where is the stress tensor and is the gravitational body force. It contains the source terms that may arise from resistances and sources. The energy equation for a fluid region can be written in terms of:

$$\frac{\partial}{\partial t}(\rho h) + \nabla\cdot(\rho h\vec{v}) = \nabla\cdot[(k + k_t)\nabla T] + S_h \tag{3}$$

where k is the molecular conductivity, k_t is the conductivity due to turbulent transport, and S_h is the source term which includes the defined volumetric heat sources. The conservation equation for species takes the following general form:

$$\frac{\partial}{\partial t}(\rho Y_i) + \nabla\cdot(\rho\vec{v}Y_i) = -\nabla\cdot\vec{J_i} + S_i \tag{4}$$

where Yi is the local mass fraction of each species, Si is the rate of creation of addition from user defined sources.

Additional transport equations were solved when the flow was turbulent. The standard *k*-epsilon transport model, which is frequently used for incompressible flows, was used to define the turbulence kinetic energy and flow dissipation rate within the model. The use of the standard *k-epsilon* transport model on building configurations was found in previous works. The turbulence kinetic energy and its rate of dissipation were obtained from the following transport equations:

$$\frac{\partial}{\partial t}(\rho k) + \frac{\partial}{\partial x_i}(\rho k u_i) = \frac{\partial}{\partial x_i}\left[\left(\mu + \frac{\mu_t}{\sigma_k}\right)\frac{\partial k}{\partial x_i}\right] + G_k + G_b - \rho\epsilon \tag{5}$$

And

$$\frac{\partial}{\partial t}(\rho\epsilon) + \frac{\partial}{\partial x_i}(\rho\epsilon u_i) = \frac{\partial}{\partial x_i}\left[\left(\mu + \frac{\mu_t}{\sigma_\epsilon}\right)\frac{\partial \epsilon}{\partial x_i}\right] + C_{1\epsilon}\frac{\epsilon}{k}\left(G_k + C_{3\epsilon}G_b\right) - C_{2\epsilon}\rho\frac{\epsilon^2}{k} \quad (6)$$

where: *Gk* represents the generation of turbulence kinetic energy due to the mean velocity gradients, *Gb* represents the generation of turbulence kinetic energy due to buoyancy. Y_M represents the contribution of fluctuating dilatation in compressible turbulence to the overall dissipation rate. C_{1e}, C_{2e} and C_{3e} are constants; σ_k and σ_e are the turbulent Prandtl numbers for k and e.

3.1.1. Computational Domain

The computational domain comprised of the building geometry, which was designed according to the actual specifications of the building, the outdoor and indoor domains, as illustrated in Figure 9. The outdoor airflow was simulated by setting one surface of the computational domain as velocity inlet and the surface located at the other end as pressure outlet. The building geometry was located at the middle of the domain and it produced a blockage of less than 10%. This was to ensure that the walls on the side of the domain had limited effect on the calculated airflow close to the building. Additionally, to further reduce this effect, the side and top walls of domain were set as symmetry.

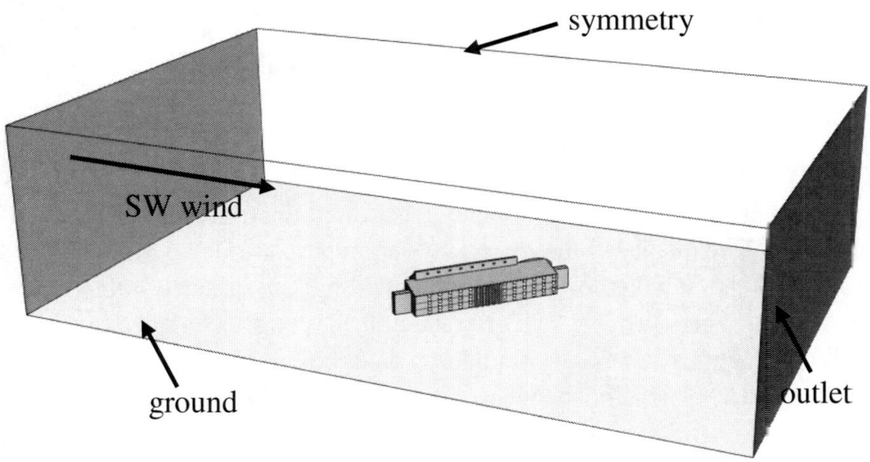

Figure 9. Computational domain.

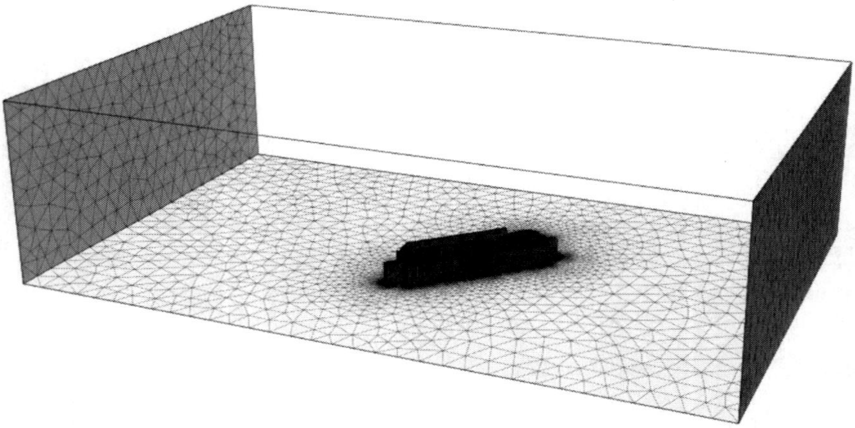

Figure 10. Mesh design.

Figure 11. (a) Mesh (b) actual building.

3.1.2. Mesh Generation

Mesh generation is one of the most important processes in a CFD simulation. The quality of the mesh plays an important role on the accuracy of results and the stability of the solution. For the investigated computational domain, patch independent CFD tetrahedron meshing technique was applied on the geometry, wherein the boundary conditions were applied on the edges and faces. The patch independent mesh algorithm for tetrahedron elements was based on the subsequent spatial subdivision algorithm, which ensured refinement of the mesh where essential, but retained larger elements where feasible, therefore allowing faster computing times. The meshed model comprised of 10,849,999 elements as displayed in Figure 10. The minimum face angle was 5.67°, while the maximum edge length and element volume ratios were 9.9 and 44.6. Figure 11 and Figure 12 display the schematic of the geometry along with the meshed model.

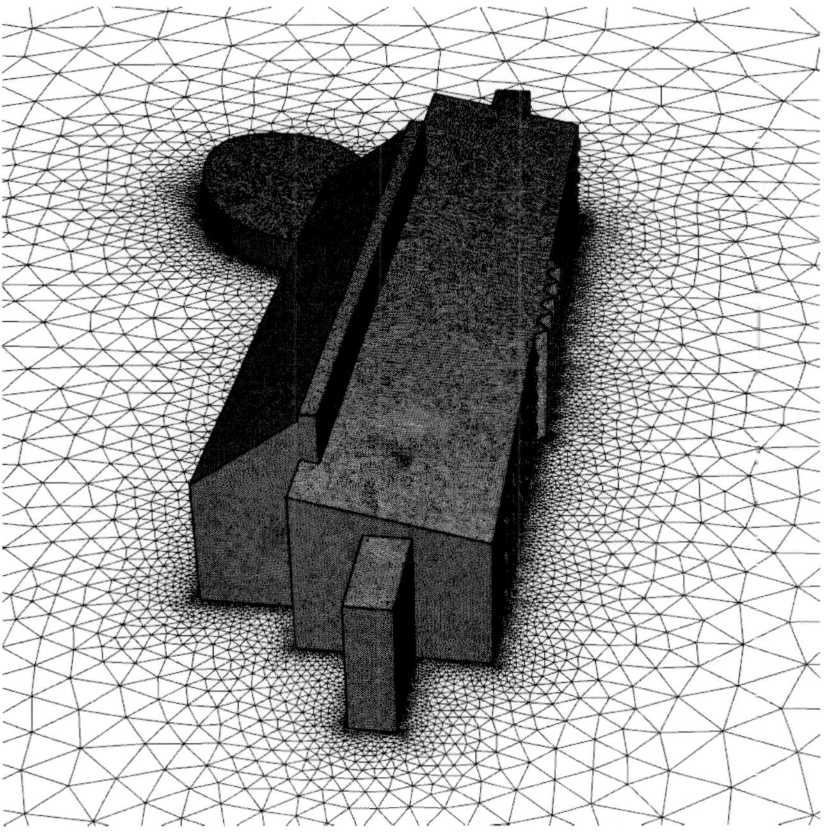

Figure 12. Mesh details.

3.1.3. Grid Sensitivity

In order to verify the accuracy of the numerical models, a grid independency test was carried out to determine the variation in results over increasing mesh sizes. Basic concepts associated with mesh refinement dealt with the refinement and evaluation of elements. Grid verification was carried out using mesh refinements (*h*-method) in order to optimise the distribution of mesh size *h* over a finite element.

The area-weighted average value of the static pressure on the three wind turbines located on the building façade was taken as the error indicator, as the grid was refined from 1,519,000 to 10,849,999 elements with the average pressure value being 4.36Pa. The grid was evaluated and refined until the posterior estimate error became insignificant between the number of elements and the posterior error indicator. The discretisation error was found to be the lowest at over ten million elements for both indicated variables and the mesh was thus selected to achieve a balance between accuracy and computational time. Figure 13 displays the variation in discretisation error at increasing number of meshed elements.

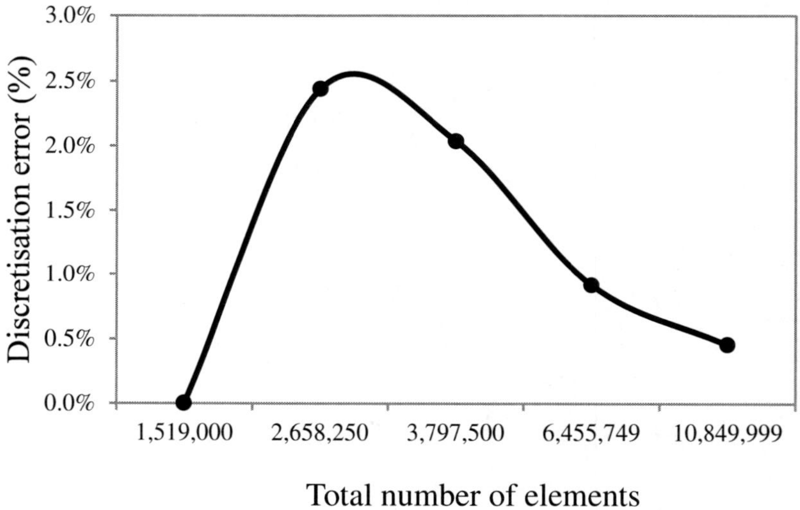

Figure 13. Mesh verification.

Table 1. Boundary conditions

Parameter	Type
Geometry	Solid zone
Enclosure	Fluid zone
Near-Wall Treatment	Standard Wall Functions
Velocity Formulation	Absolute
Solver Type	Pressure-Based
Time	Steady
Gravity	-9.81 m/s^2
Velocity Inlet	5 m/s at SW direction
Temperature Inlet	291 K
Humidity Inlet	71 %
Pressure Outlet	Atmospheric

3.1.4. Boundary Conditions

The applied boundary conditions comprised of a reference velocity of 5m/s approaching SW to the building, as presented in Table 1. The geometry was modelled as a solid zone, while the indoor-outdoor domain was modelled as a fluid zone for the analysis. The boundary conditions were kept identical throughout the numerical investigation for all analysed models.

In order to simulate the internal gain of the people, lighting and equipment, the floor surface was set as heat flux with a heat flux value of 33.7 W/m^2.

Table 2. CIBSE Guide internal gain for a typical office building

Building type	Use	Density of occupation / person·m^{-2}	Sensible heat gain / W·m^{-2}			Latent heat gain / W·m^{-2}	
			People	Lighting*	Equip't†	People	Other
Offices	General	12	6.7	8–12	15	5	—
		16	5	8–12	12	4	—
	City centre	6	13.5	8–12	25	10	—
		10	8	8–12	18	6	—
	Trading/dealing	5	16	12–15	40+	12	—
	Call centre floor	5	16	8–12	60	12	—
	Meeting/conference	3	27	10–20	5	20	—
	IT rack rooms	0	0	8–12	200	0	—

3.2. Building Thermal Modelling

In this study, the Ecotect software was used as a research tool and setting regional climate conditions as a starting point, the authors simulated and analysed the building orientation, daylight in building of the commercial building. Ecotect software is an environmental analysis tool that allows designers to simulate building performance right in conceptual phase. It includes a wide array of detailed analysis functions with a highly visual and interactive display that makes analytical results can be directly presented within the context of the building model. Thus, complex concepts and extensive datasets can be communicated in surprisingly intuitive and effective ways.

3.2.1. Weather Data

The case study building is located in Scotland, which is at 55° 57′ north latitude and 3° 11‴ east longitude. The Scotland climatic data was inserted into weather tool of Ecotect, which output the analysis results of the climate. Scotland has a temperate climate, and the weather changes constantly. Figure 14 and Figure 15 are respectively the distribution of annual hourly dry-bulb temperature and the distribution of annual hourly humidity generated by Ecotect Weather Tool. Figure 14 shows that Scotland has a typical mild summer and cold winter climate. The average annual dry-bulb temperature is about 8 °C. The coldest month is January and the average temperature is 2.8°C, while the hottest month is July and the average temperature is 14.5 °C. The total cooling degree hours is 81 hours and the total heating degree hours is 83,149 hours. Figure 15 shows that the annual humidity is Scotland is high. The average relative annual humidity is between 80–90%. These climatic conditions are mainly outside the thermal comfort zone specifically during the winter. To ensure an acceptable indoor thermal condition, careful consideration is given to insulation in an energy-saving design.

Scotland is extremely windy, due to eastward moving Atlantic that brings strong winds continuously throughout the year. The wind prevails from south-west and the annual mean wind speed is 5 m/s, as presented in Figure 16.

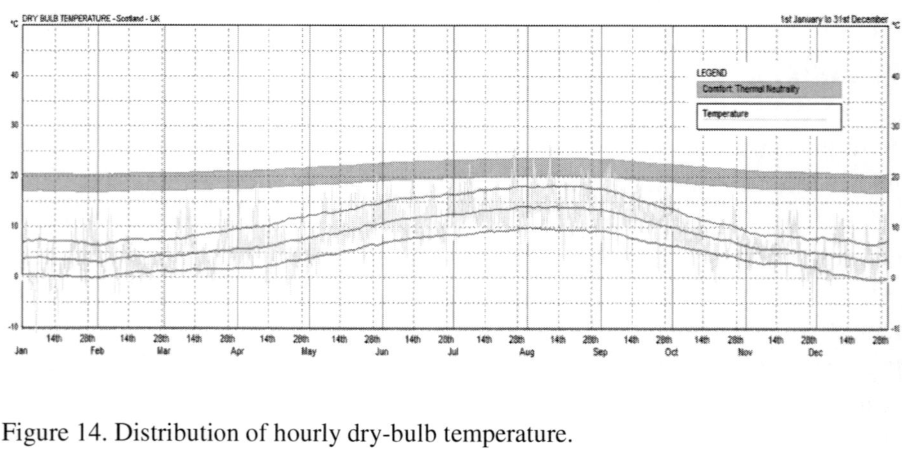

Figure 14. Distribution of hourly dry-bulb temperature.

3.2.2. Geometry

A full scale geometry of the building was modelled for thermal analysis using the Autodesk Ecotect software. The building consisted of three floors and each floor was divided into zones to apply different occupancy schedule and activity, internal gain profiles and infiltration rate. Figure 17 and Figure 18 show the full geometry of the building extracted from the CAD drawing.

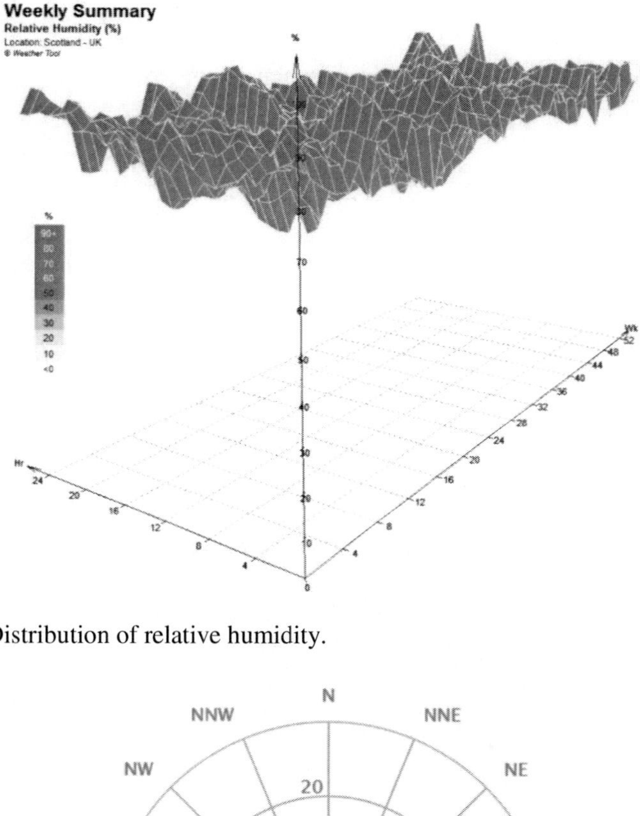

Figure 15. Distribution of relative humidity.

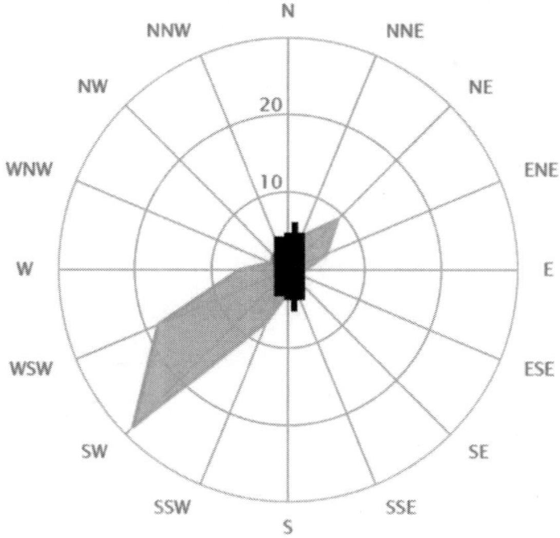

Figure 16. Distribution of wind around the building.

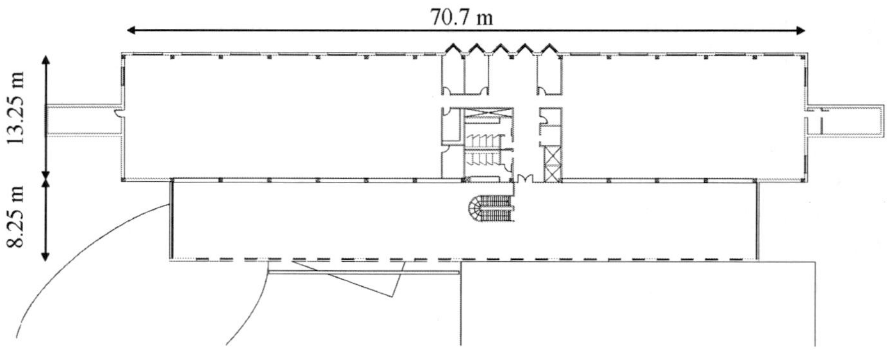

Figure 17. Floor plan of the studied building used for the Ecotect design modeller.

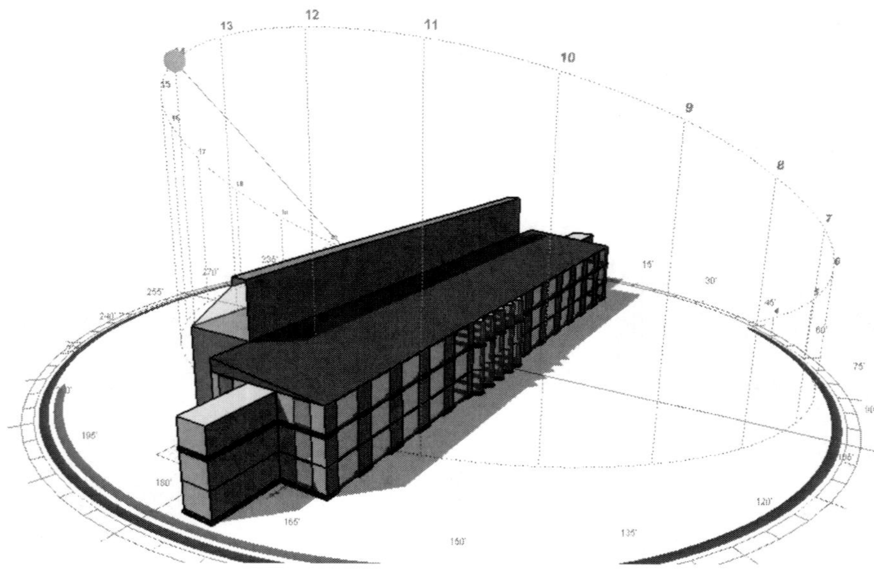

Figure 18. Ecotect thermal model showing the summer sun path.

Table 3. Summary of building materials for the BES model

Parameter	Materials	U-Value (W/m^{2}k)
External walls	Concrete block plaster	1.80
Internal walls	Concrete block plaster	1.80
External windows	Double glazed	2.70
Flat roof	Concrete roof	0.90
Ground	Concrete slab on ground	0.88

3.2.3. Thermal Properties

Table 4, Table 5 and Table 6 displays the thermal properties, internal design conditions and occupancy schedules used in the modeling. The natural ventilation systems was selected to operate for 12 hours on weekdays and the comfort band was set between 18 °C - 26°C.

Table 4. Thermal properties

Parameters	Set Values
HVAC	Mixed-Mode System
Comfort band	18°C - 26 °C
Hours of operation	7.00 AM to 7:00 PM (Weekdays) Weekends Off

Table 5. Internal design conditions

Parameters	Set Values
Clothing	1 clo (light suit)
Lighting level	400 Lux (office)
Humidity	50 %
Air speed	0.5 m/s
Infiltration rate	0.5 ach (average)

3.3. Field Study

Field studies of thermal comfort were applied in this study for a week in August 2012. Occupants' thermal sensation and preference were measured using survey questionnaires based on the ASHRAE seven-point scale, as presented in Table 7.

Table 6. Occupants

Parameters	Set Values
Number of Occupants	125 (ground floor) 125 (first floor) 125 (second floor) 20 (atrium)
Occupant activity	70W - sedentary (offices) 80W - walking (atrium)
Schedule	9.00 AM to 5:00 PM (Weekdays)

Table 7. Thermal sensation and preference questions based on the ASHRAE seven-point scale

Currently at my desk, regarding the thermal environment I feel:							
Hot	Warm	Slightly warm	Neutral	Slightly cool	Cool	Cold	No strong opinion
+3	+2	+1	0	-1	-2	-3	
Currently at my desk, I prefer the thermal environment was:							
Much warmer	Warmer	Slightly warmer	No change	Slightly cooler	Cooler	Much cooler	No strong opinion
+3	+2	+1	0	-1	-2	-3	

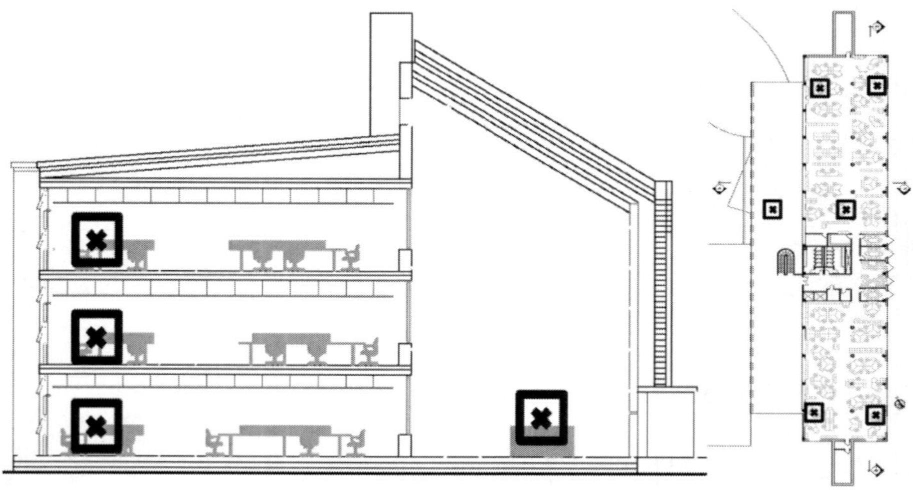

Figure 19. Measuring points for constant measurements.

Environmental measurements were applied instantly at the workstation level simultaneous with the occupant responding to the questionnaire and constantly at particular measuring points around the building to record the overall performance of the building. The constant measuring points were selected around the building on the floor, desk and ceiling levels, as demonstrated in Figure 19.

Humidity, temperature and air monitoring equipment were used to measure the thermal environment, as presented in Table 8. Mean radiant temperature was calculated using surface measurements using the constant measuring method.

Table 8. Equipment for environmental measurements

Measurement	Time	Equipment details	Resolution	Accuracy	Range
Dry bulb temperature	Instant: at workstations	PCE-GA 70 air quality meter	0.1°C	±0.5°C	5 to 50°C
Relative humidity	Instant: at workstations	PCE-GA 70 air quality meter	0.1°C	±3 RH	10 to 90% RH
Carbon dioxide level	Instant: at workstations	PCE-GA 70 air quality meter	1 ppm	±50 ppm	6000 ppm
Dry bulb temperature	Constant: set in particular locations	Tiny Tag Plus 2 TGP-4500	0.01°C	0.01°C	-25 to +85°C
Relative humidity	Constant: set in particular locations	Tiny Tag Plus 2 TGP-4500	0.3% RH	±3% RH	0 to 100% RH

4. RESULTS

4.1. Computational Fluid Dynamics (CFD)

Using the building model, Figure 20 displays the static pressure distribution on the surfaces of the building geometry and the ground. As observed, the positive pressure was predicted at the south-west wall of building. This was due to the force being directly perpendicular to the area of interaction. As a result, a negative pressure was created on the opposite end at the immediate downstream of the building, as presented in Figure 21. The window openings located on the negative pressure side caused the airflow to be pulled out from the building and extracted the old air from the occupied spaces. The maximum positive pressure was estimated at 14.2 Pa, while the maximum negative pressure was observed to be -17.5 Pa. The analysis showed that a maximum pressure differential of 31.7 Pa was created for the entire building geometry.

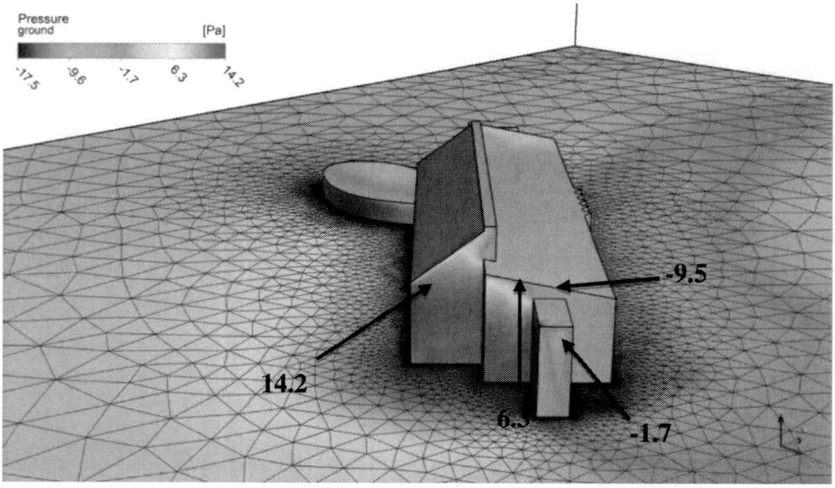

Figure 20. Surface pressure analysis of the south wall of building.

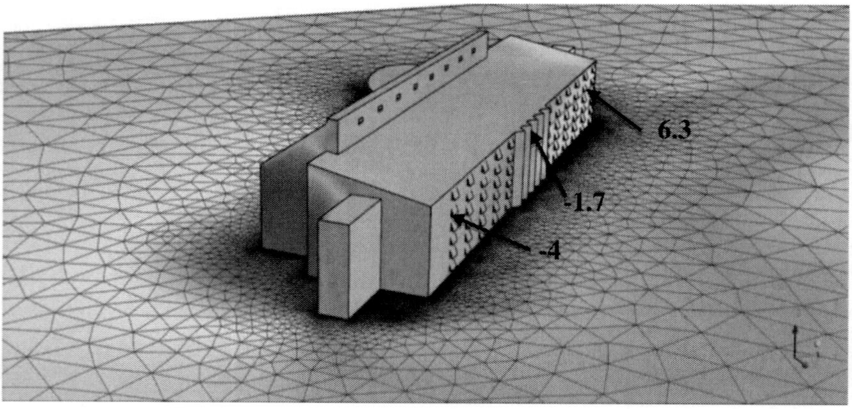

Figure 21. Surface pressure analysis of the east wall of building.

Figure 22 displays the velocity streamline around the building model. As seen in the diagram, the airflow coming from the left (south-west of the building) reduced speed as it approached the wall of the building and accelerated and cavitated as it moved along the roof and corner of the geometry. This created strong negative pressure at the corners and less strong negative pressure on the rest of the building walls and roof. From the streamline analysis, a large vortex was observed downstream of the airflow path.

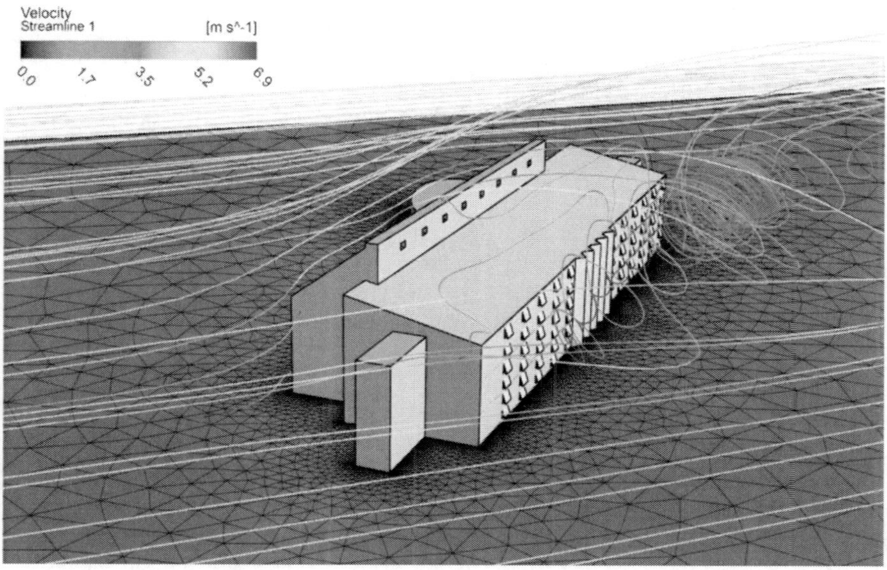

Figure 22. Velocity streamlines around the building model.

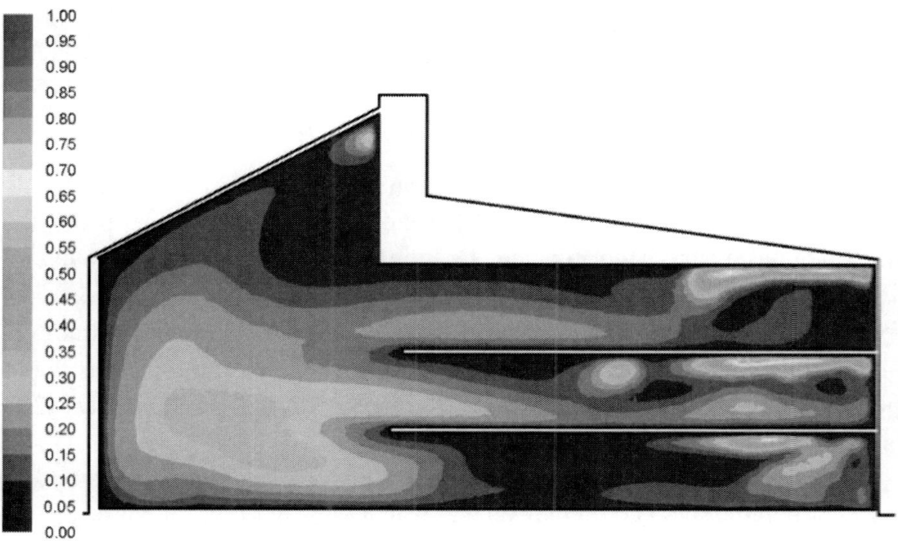

Figure 23. Indoor velocity contour.

Figure 23 shows a cross-sectional contour plot of the air velocity distribution inside the building model. As observed, the highest airflow speed (up to 1 m/s) was calculated near the windows, where airflow was extracted.

On average, the airflow speed inside the building was between 0.1 and 0.4 m/s. Figure 24 shows the pressure distribution inside the building. The average indoor pressure was -4.8 Pa. A ventilated space under negative pressure meant that more airflow was being removed than delivered, which was the case for the building when the wind was from south-west.

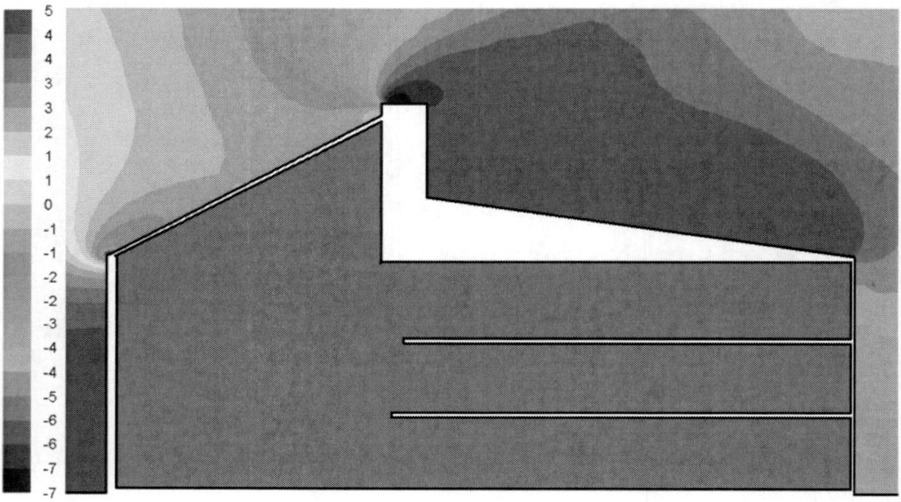

Figure 24. Indoor static pressure contour of a cross-sectional plot.

Figure 25 depicts the predicted temperature distribution inside and outside the building. The heat flux boundary condition was used to replicate the internal gains from the occupants, lighting and equipment. This boundary condition was applied across floor surface and was set to 33.7 W/m^2. As presented in Figure 25, higher temperature can be observed in areas with lower air speed and lower temperature in areas with higher air drafts, such as the in the ground floor and the atrium. The average indoor temperature was 22°C when the outdoor temperature was set at 18°C.

Figure 26 illustrates the predicted relative humidity distribution inside and outside the building. The humidity level was inversely proportional to the air temperature. The humidity level was lower in areas with higher temperature and humidity level was higher in areas with lower temperature. The average indoor relative humidity was 57% when the outdoor relative humidity was set to 71%.

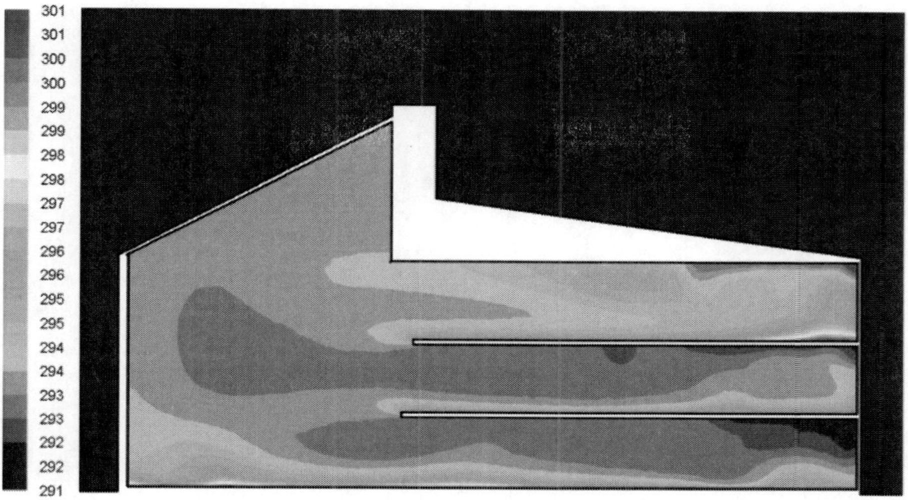

Figure 25. Indoor temperature contour of a cross-sectional plot.

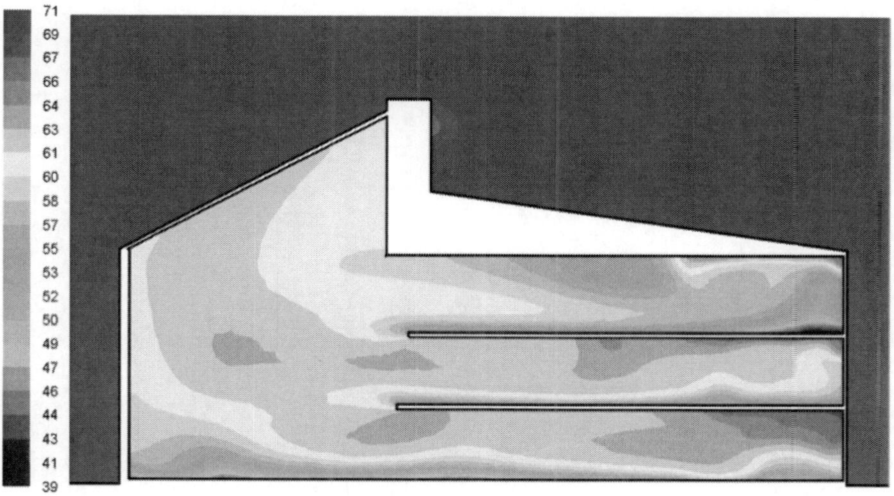

Figure 26. Indoor relative humidity contour of a cross-sectional plot.

4.2. Building Thermal Modelling (BTM)

Figure 27 and Figure 28 show the hourly temperature inside the ground floor during a typical summer day (August 18 and 21 respectively). As observed, the temperature inside the ground floor was between 21 – 26 °C during occupation time when the outdoor temperature was in the range of 15 - 18 °C.

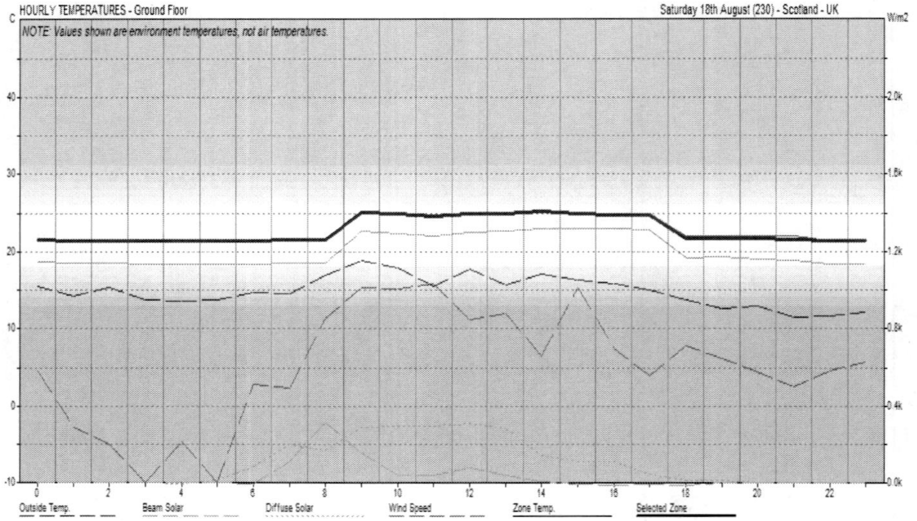

Figure 27. Hourly temperature profile in the ground floor during the month of August (18th).

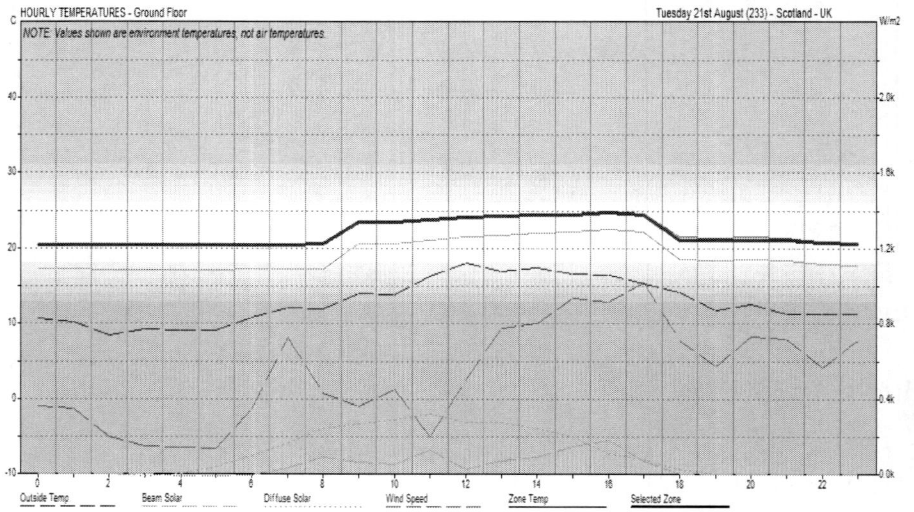

Figure 28. Hourly temperature profile in the ground floor during the month of August (21st).

The ECOTECT software allows the calculation of the dynamic spatial comfort for a thermal zone using the analysis grid. The analysis grid can be used to cover the floor area of the model. Figure 29 shows the calculated Predicted Mean Vote (PMV) values mapped over the analysis grid in the XY axis, displaying comfort values horizontally. As observed, the calculated PMV value for the entire floor is -0.36. Based on the ASHRAE sensation scale in Table 7, this value is between "Neutral" and "Slightly Cool", which indicates that the comfort conditions inside the space are within the acceptable range. Figure 30 shows the calculated Predicted Percentage Dissatisfied (PPD) values mapped over the analysis grid. Based on the analysis grid, about 8.48% - 8.94% were dissatisfied with their comfort conditions.

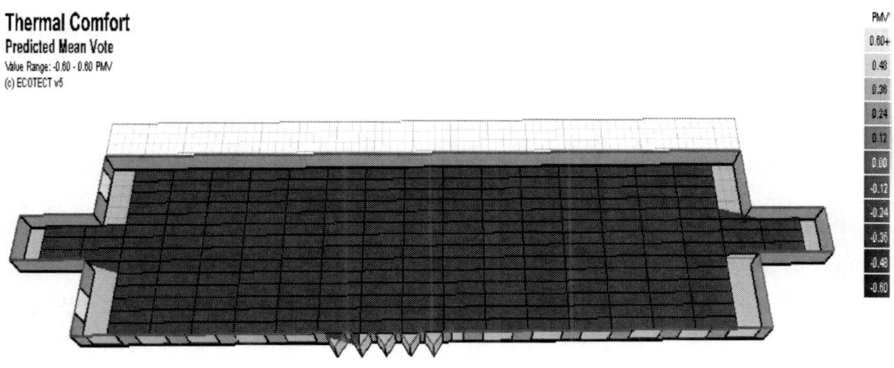

Figure 29. Calculated Predicted Mean Vote (PMV) during August 21.

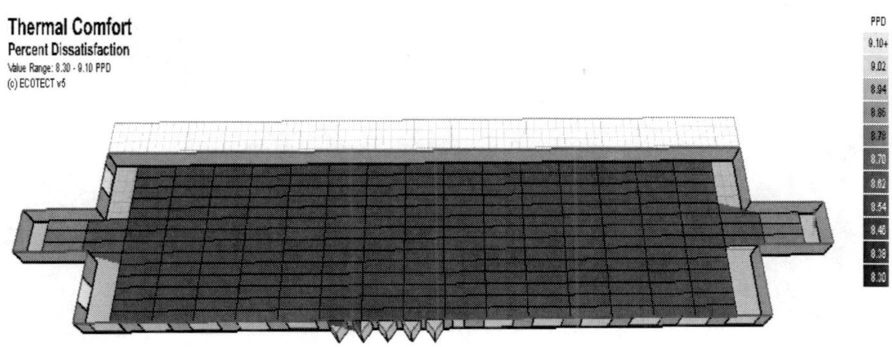

Figure 30. Calculated Percentage Dissatisfaction (PPD) during August 21.

Figure 31 and 32 compares the analysis of the interior daylighting during summer and winter. This tool can be used to quantify the amount of natural daylighting, which can help create comfortable space, reduce artificial lighting loads and reduce the cooling loads.

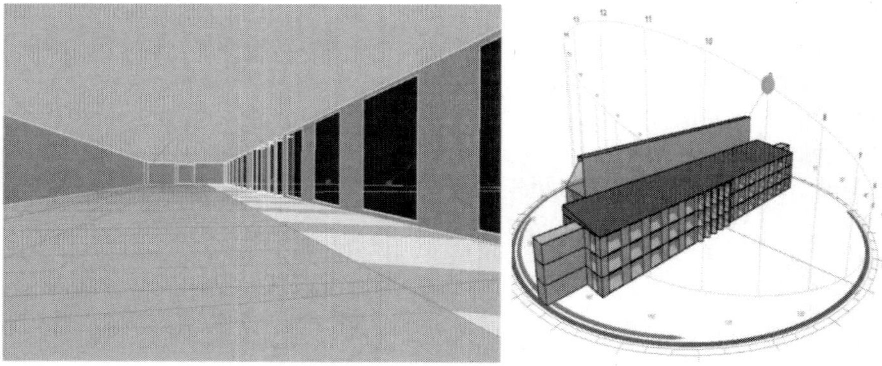

Figure 31. Comparison of natural daylight during the summer (July) at 9 AM.

In order to see where the building's heating and cooling loads come from one should look at where the building is gaining and losing heat. The passive gains breakdown chart in Figure 33 shows what percentage of the gains and losses were coming from conduction, solar radiation, infiltration, and internal loads.

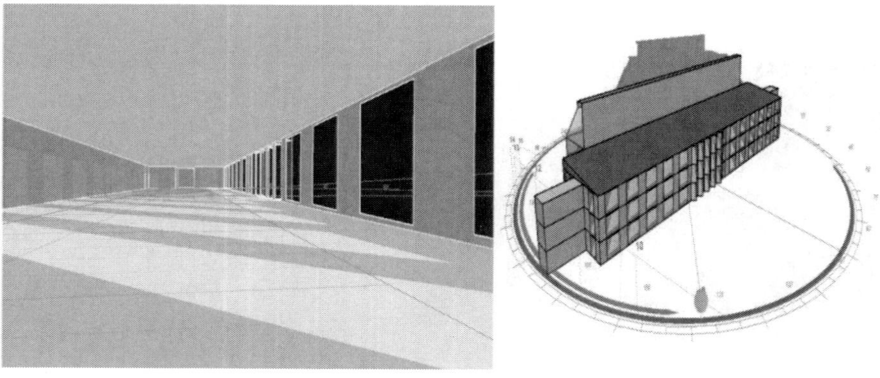

Figure 32. Comparison of natural daylight during the winter (January) at 9 AM.

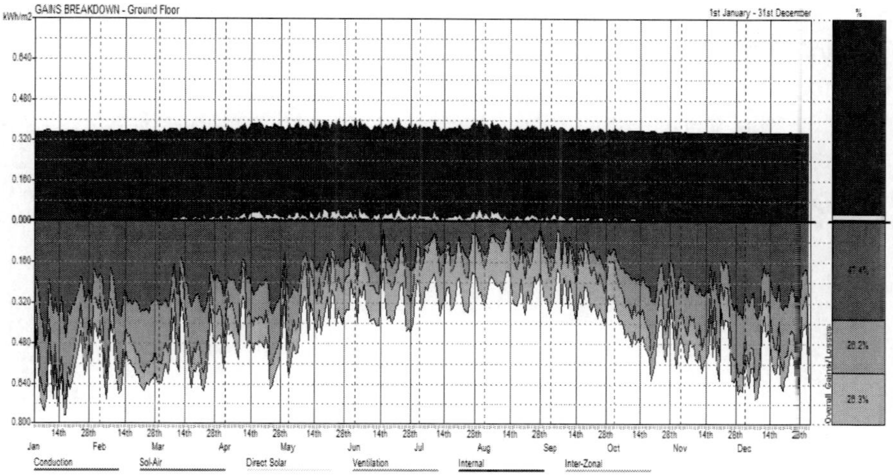

Figure 33. Passive gains breakdown.

4.3. Field Study

The energy performance of the building was compared against the CIBSE benchmark (CIBSE, 2003), as presented in Figure 34. The energy data was provided by the building management. The energy use of the building was 550 Kwh/m^2 per year, which was well within the acceptable range and very close to the energy consumption of a good practice of a naturally ventilated open plan office. This indicated that the application of energy strategies and natural ventilation was successful in reducing the energy demand of the building.

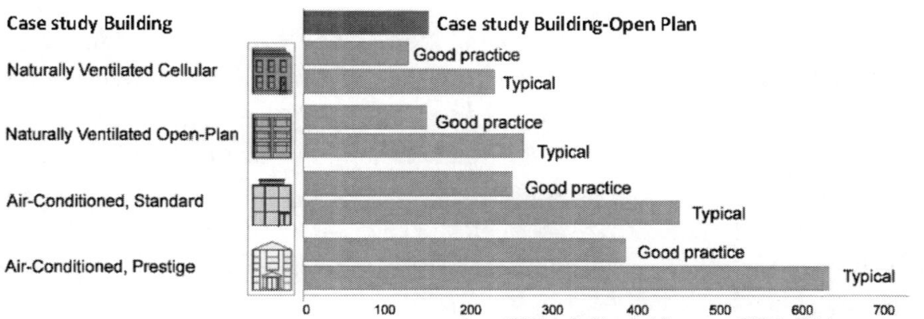

Figure 34. Energy consumption of the building comparing the buildings against the benchmark (CIBSE, 2003), KWh/m^2 per year.

Thermal performance of the building was compared against the ASHRAE Standard 55-2013, as presented in Figure 35. This was based on the environmental measurements at workstations and each dot in Figure 35 represents the thermal condition of a particular workstation. This analysis showed that although thermal conditions are within the acceptable range, the majority of them fall into the winter conditions estimating a slightly cool thermal environment.

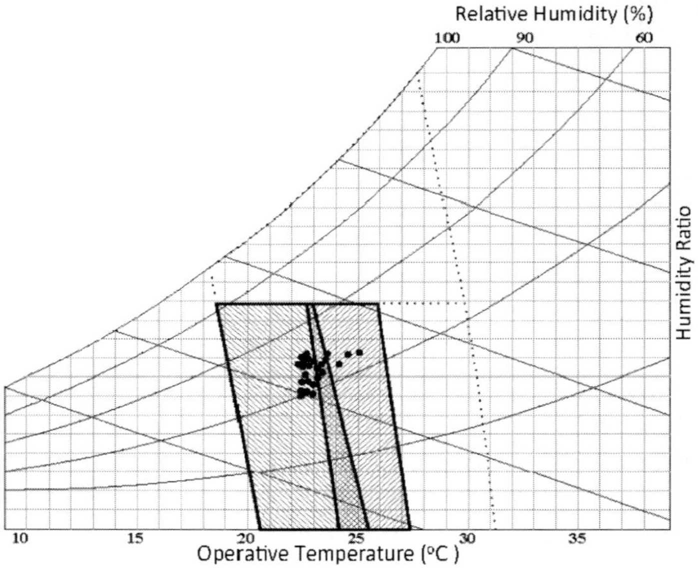

Figure 35. Thermal performance of the measured workstations according to the ASHRAE Standard 55-2013.

This agreed with the results of simulation, which was illustrated in Figure 25, Figure 27 and Figure 28. The results of simulation showed indoor temperatures between 20-26˚C, which was similar to the collected data from the site, as presented in Figure 35. Relative humidity of the simulation also agreed with that of the field study, respectively Figure 26 and Figure 35, which were both between 50-60% RH. The field data regarding the air movement in the office showed less than 0.1 m/s air movement in the majority of the workstations, which agrees with the results of the simulation, as presented in Figure 23. Occupants view of thermal environment was analysed based on the ASHRAE seven point scale thermal sensation and thermal intention, as discussed in 3.3. The analysis showed that about 40% of the occupants feel neutral (i.e. 35%) and want no change in the temperature (i.e. 42%), as presented in Figure 36, which is the bar chart of this analysis.

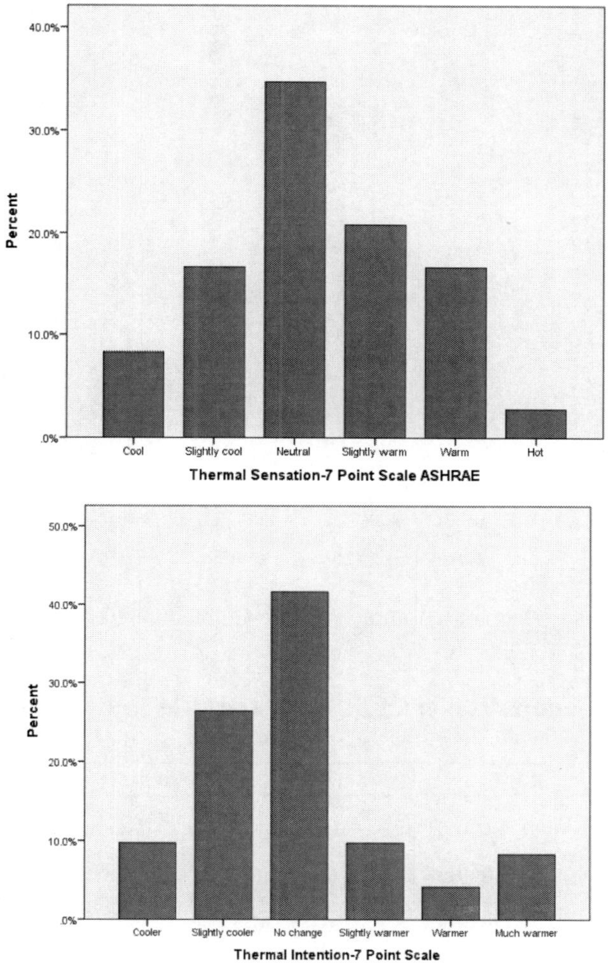

Figure 36. Occupants' response based on the surveyed ASHRAE seven point scale thermal sensation and thermal intention.

Figure 37 shows the boxplot of thermal sensation, thermal intention and the expected response. It shows that majority of responses are within the expected range, although some respondents fall out of this range. For instance, some of the respondents who feel cool wanted no change in the temperature.

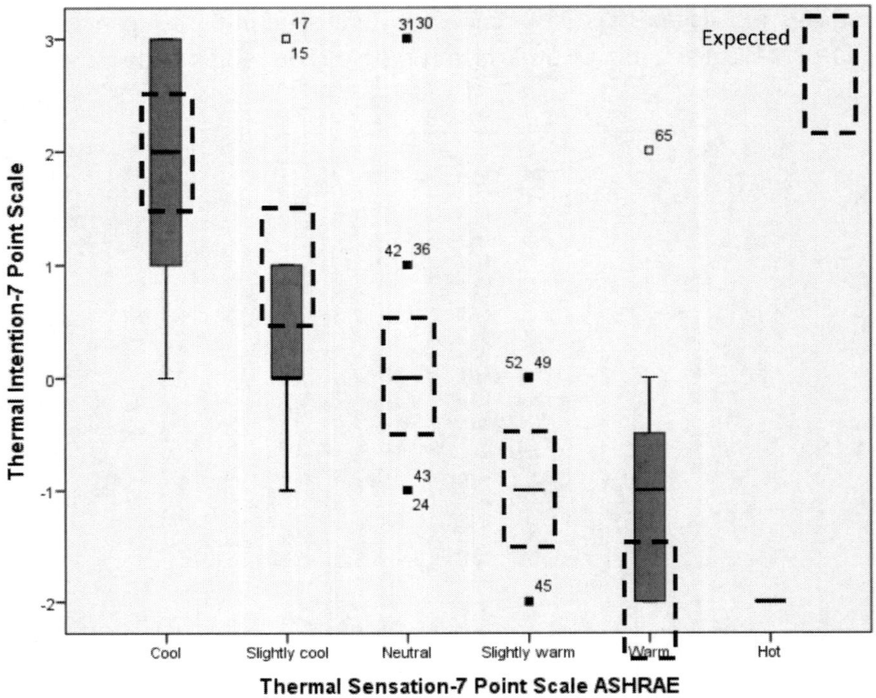

Figure 37. Boxplot of thermal sensation and thermal intention of occupants based on the ASHRAE seven point scale.

Table 9. Comparison of CFD, BTM and field study PMV values

Method	Point 1	Point 2	Point 3	Point 4
CFD	-0.01(Neutral)	-0.41 (Neutral)	-0.45 (Neutral)	-0.55 (Slightly cool)
BTM	-0.31 (Neutral)	-0.29(Neutral)	-0.36(Neutral)	-0.35(Neutral)
Field Study	-0.33 (Neutral)	-0.61 (Slightly cool)	-0.69 (Slightly cool)	0 (Neutral)

The PMV analysis based on the environmental measurements showed that 35% of the occupants were expected to feel neutral and the rest were expected to feel slightly cool. This was in line with the findings of thermal performance illustrated in Figure 35. The number of respondents expected to have a neutral thermal sensation agreed with the number of occupants' responses, who actually had a neutral thermal sensation (i.e. 35%). These findings agreed with the findings of the simulation, as similar thermal conditions were found. Based on the BTM analysis, occupants were expected to feel between "neutral" and "slightly cool" (PMV = -0.36), as discussed in section 4.1. The PPD based on the field data was 14.9%, which agreed with the results of the BTM indicating 8.48% - 8.94% Predicted Percentage Dissatisfied, as discussed in section 4.1.

Points in Table 9 were selected from Figure 19: Point 1 (second floor), Point 2 (first floor), Point 3 (ground floor) and Point 4 (atrium.

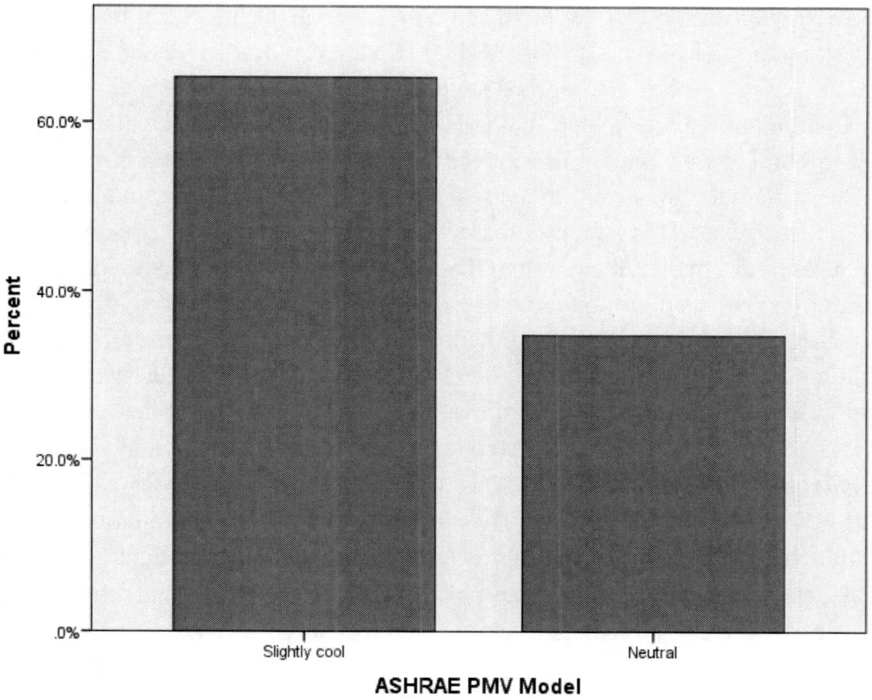

Figure 38. PMV model based on the environmental measurements.

Conclusion

The results indicated that the indoor air temperature, relative humidity and air velocity of the simulations agreed with the results of the field data. The simulation showed temperatures between 20-26°C, which was very close to the temperature in the actual building. The PMV model based on the field data showed that 35% of the occupants were expected to feel neutral and the majority of the occupants were expected to feel slightly cool, as these temperatures fell in the winter section of thermal comfort zone. Based on the survey, 35% of the occupants reported to have a neutral thermal sensation. This agreed with the results of the simulation, as occupants were expected to feel between "neutral" and "slightly cool" (PMV = -0.36). The detailed analysis of PMV based on the floor areas also revealed close agreement between the three different methods of CFD, BTM and Field Study. The PPD analysis based on the field data indicated that 14.9% of the occupants were dissatisfied with the thermal environment. This was in line with the results of the simulation (PPD = 8.48% - 8.94%). Relative humidity of the simulation was 57% RH, which also agreed with the field data (i.e. 50-60% RH). The CFD simulation also showed that most of the office had sufficient ventilation lower than 0.1 m/s, which also agreed with the field data. Overall, the results of the two simulations were in agreement with the results of the analysis based on the field data. This suggests that simulation tools, such as CFD and BTM, are useful to predict the thermal behavior of the building and occupants' comfort as well as to improve the overall design of the building.

Natural ventilation has the potential to improve the energy efficiency of buildings by reducing reliance on air-conditioning. To effectively design buildings with natural ventilation, evaluation tools are needed that are able to assess its performance. This study used Computational Fluid Dynamics (CFD), Building Thermal Modelling (BTM) and field study to evaluate the performance of an open plan office in in which natural ventilation was in operation. Through the integration of CFD and BTM simulations, this work evaluates the quality of the thermal environment of an open plan office. Because of the complementary nature of BTM and CFD results, an integration of both methods can eliminate many assumptions of the simulation, which increases the accuracy. The BTM requires detailed information on the airflow and thermal conditions of the surrounding of the building. These can be computed in the CFD simulation to improve the accuracy of the prediction of the indoor thermal environment. The results indicate that BES and CFD can play an important role to improve the design of the building to achieve this balance. They provide important information to optimise the energy efficiency and the quality of the thermal environment.

ACKNOWLEDGMENT

The authors gratefully acknowledge the contribution of the architect, Donald Canavan, as well as the management and occupants of the case study building.

References

ASHRAE 2009. *ASHRAE Handbook: Fundamentals. SI Edition ed. Atlanta: American Society of Heating,* Refrigerating and Air-Conditioning Engineers, Inc.

ASHRAE 2010. ASHRAE Standard 55. In: COOPER & W., K. (eds.) *Thermal Environmental Conditions for Human Occupancy. Atlanta: American Society of Heating,* Refrigerating and Air-Conditioning Engineers.

ASHRAE 2013. ANSI/ASHRAE Standard 55-2013: *Thermal Environmental Conditions for Human Occupancy.* Atlanta: American Society of Heating, Refrigerating and Air-Conditioning Engineers, Inc.

Auliciems, A. & Szokolay, S., V. 1997. Thermal Comfort. PLEA: *Passive and Low Energy Architecture International, Design Tools and Techniquees.* Note 3. In Association wiht the University of Queensland Dept. of Architecture.

Bedford, T. 1936. The Warmth Factor in Comfort at Work: A Physiological Study of Heating and Ventilation. *MRC Industrial Health Research Board Report.* London: H.M.S.O.

Bromley, D., Basil 1990. Academic Contributions to Psychological Counselling 1. A Philosophy of Science for the Study of Individual Cases. *Counselling Psychology Quarterly,* 3, 299-307.

Calautit, J. K., Hughes, B. R. & Ghani, S. A. 2013a. A Numerical Investigation into the Feasibility of Integrating Green Building Technologies into Row Houses in the Middle East, *Architectural Science Review*, 56, 279-296.

Calautit, J. K., Hughes, B. R. & Ghani, S. A. 2013b. Numerical investigation of the integration of heat transfer devices into wind towers, *Chemical Engineering Transactions*, 34, 43-48.

Calautit, J. K., Chaudhry, H.N., Hughes, B. R. & Ghani, S. A. 2013c. Comparison between evaporative cooling and heat pipe assisted thermal loop for a commercial wind tower in hot and dry climatic condition, *Applied Energy,* 101, 740-755.

Calautit, J. K., Hughes, B. R., Chaudhry, H.N. & Ghani, S. A. 2013d. CFD analysis of a heat transfer device integrated wind tower system for hot and dry climate, *Applied Energy,* 112, 576-591.

Calautit, J. K. & Hughes, B. R. 2014a. Wind tunnel and CFD study of the natural ventilation performance of a commercial multi-directional wind tower, *Building and Environment*, 80, 71-83.

Calautit, J. K. & Hughes, B. R 2014b. Measurement and prediction of the indoor airflow in a room ventilated with a commercial wind tower, *Energy and Buildings*, 84, 367-377.

Calautit, J. K., O'Connor, D. & Hughes, B. R. 2014c. Determining the optimum spacing and arrangement for commercial wind towers for ventilation performance, *Building and Environment*, 82, 274-287.

Chen, Y., Liu, J., Pei, J., Cao, X., Chen, Q. & Jiang, Y. 2014. Experimental and simulation study on the performance of daylighting in an industrial building and its energy saving potential, *Energy and Buildings,* Volume 73: April: 184-191.

Cibse 2003. Energy Consumption Guide 19: Energy Use in Offices. Best Practice Programme. *The Chartered Institution of Building Services Engineers.*

Etheridge, D. 2011. *Natural ventilation of buildings: theory, measurement and design.* Wiley, UK.

Fidel, R. 1984. The Case Study Method: A Case Study. *LISR,* 6, 273-288.

Hughes, B. R., Calautit, J. K. & Ghani, S. A. 2012. The Development of Commercial Wind Towers for Natural Ventilation: a review. *Applied Energy*, 92, 606-627.

Hughes, B. R., Chaudhry, H.N. & Calautit, J. K. 2012. Passive energy recovery from natural ventilation air streams. *Applied Energy*, 113, 127-140.

Humphreys, M. A. & Nicol, J. F. 1995. An Adaptive Guideline for UK Office Temperatures; UK Adaptive Temperature Guideline. In: Nicol, J. F., Humphreys, M. A., Sykes, O. & Roaf, S. (eds.) *Standards for Thermal Comfort: Indoor Air Temperatures for the 21st Century.* London: E & FN Spon.

Humphreys, M. A. 1977. Clothing and the Outdoor Microclimate in Summer. *Building and Environment,* 12, 137-142.

Humphreys, M. A. 1978. Outdoor Temperatures and Comfort Indoors. *Building Research and Practice,* 6, 92-105.

Johansson, R. 2003. Case Study Methodology. Methodologies in Housing Research. *Stockholm: Royal Institute of Technology and International Association of People-Environment Studies.*

Jones-Lee, M. & Loomes, G. 2003. Valuation of safety, D.A. Hensher, K.A. Button (Eds.), *Handbook of Transport and the Environment,* Elsevier Science, Oxford.

Nicol, F. & Roaf, S. 2005. Post Occupancy Evaluation and Field Studies of Thermal Comfort. *Building Research and Information,* 33, 338-346.

Nicol, F., Humphreys, M. A. & Roaf, S. 2012. *Adaptive Thermal Comfort: Principles and Practice, Routledge.*

Nicol, J. F. & Humphreys, M. A. 1973b. Thermal comfort as part of a self-regulating system. *Builiding Research & Information,* 1, 174-179.

Nicol, J. F. 2004. *Post occupancy evaluation and field studies of thermal comfort. Closing the Loop.* Windsor: NCEUB.

Nielsen, PV. 2004. Computational fluid dynamics and room air movement. *Indoor Air,* 14 (Suppl. 7), 134–143.

O'Connor, D.,Calautit, J. K., & Hughes, B. R.. 2014. A Study of Passive Ventilation Integrated with Heat Recovery, *Energy and Buildings*, 82, 799-811.

Sofotasiou, P., Hughes, B. R., & Calautit, J. K. 2015. Qatar 2022: Facing the FIFA World Cup climatic and legacy challenges, *Sustainable Cities and Society*, 14, 16-30.

Schueter, A. & Thessling, F. 2008. Building information model based energy/exergy performance assessment in early design stage. *J Autom Constr.*

Shi, X. & Yang, W. 2013. Performance-driven architectural design and optimization technique from a perspective of architects, *Automation in Construction,* Volume 32: July: 125-135.

Short, C. A., Lomas, K.J. & Woods, A. 2004. Design strategy for low-energy ventilation and cooling within an urban heat island *Build Res Inf,* 32 (3): 187–206.

Yang, L., HE, B. & YE, M. 2014. Application research of Ecotect in residential estate planning, *Energy and Buildings,* Volume 72: April: 195-202.

In: Natural Ventilation
Editor: Oscar T. Haynes

ISBN: 978-1-63482-414-9

Chapter 3

COMMERCIAL WIND TOWERS FOR HEATING AND COOLING BASED ON CLIMATIC CONDITIONS

Dominic O'Connor*, John K. Calautit and Ben Richard Hughes
Energy Engineering Group, Department of Mechanical Engineering, University of Sheffield, UK

ABSTRACT

Commercial wind towers are passive ventilation technology based on traditional vernacular wind towers of the Middle East, particularly in climates of hot, arid conditions. By manipulating pressure differences and the buoyancy effect, caused by wind driven flow and density stratification of internal air due to temperature differences, adequate levels of ventilation can be provided to buildings utilising wind towers in a low energy process. This negates the use of mechanical air-conditioning systems which require substantial amounts of energy. One current limitation of commercial wind towers which exists is the inability to condition the incoming air for thermal comfort to occupants. This relates to cooling demand in hot climates and heating and cooling demand in temperate climates. At present, commonly used methods to provide thermal comfort rely on energy intensive mechanical processes to cool or

* Corresponding author: Email: dboconnor1@sheffield.ac.uk.

heat the incoming air which is not sustainable given pressure to reduce energy use and global greenhouse gas emissions.

Low energy methods for cooling and heating of air passing through a wind tower have not yet been comprehensively explored. A cooling system utilising the combination of heat pipes and a heat sink has been developed and incorporated into a commercial wind tower. This system has been analysed using CFD and tested experimentally in a hot climate and showed positive results in reducing the temperature of the incoming air. A heat recovery system utilising a rotary thermal wheel to transfer the thermal energy from the exhaust airstream of a wind tower has been developed and incorporated into a four-sided commercial wind tower.

INTRODUCTION

Driven by an ever increasing global demand for energy across all aspects of life and industry, greenhouse gas emissions have increased at an alarming rate. Furthermore, the development of emerging economies has led to significant addition to greenhouse gas emissions that had not previously been taken into account. As a means to counter the rising greenhouse gas emissions, a strong drive for more energy efficient technologies has gained popularity amongst consumers, whilst governments have been bound to statutory requirements to cut greenhouse gas emissions from pre-1990 levels by 80% by the year 2050[1]. To achieve this ambitious reduction in emissions is not unfeasible but requires a societal movement away from energy intensive processes and a move toward low energy and zero energy technology.

The building sector has substantial scope to reduce the energy use associated with the operation and maintenance of buildings worldwide. Presently, the building sector contributes up to 40% of the global energy demand [2, 3]. Of this total energy demand heating, ventilation and air-conditioning, commonly referred to as HVAC, accounts for 60% [4-6]. This is due to the thermal comfort demands of occupants, regulations for adequate ventilation supply rates and current HVAC technology that is commonly used.

Addressing the substantial energy requirements of mechanical HVAC services has the potential to significantly reduce the energy demands of buildings heavily reliant on such systems such as commercial buildings, schools and office spaces. A notable portion of the energy demand arises from the use of electrical fans to circulate air around the ductwork to deliver air to the appropriate spaces. Using low energy fans which are capable of supplying the same volume of air is a preferable solution to such a problem and is becoming a widespread solution. Though reducing energy demand through the use of low energy fans is one solution in cutting greenhouse gases as a result of HVAC systems, the key area for reduction is the ventilation and conditioning of the air to the correct temperature.

Outdoor air should be constantly supplied to a building to ventilate of air that can become polluted due to the presence and activity of occupants. Ideally, this ventilation is controlled by systems put in place by designers and engineers. Uncontrolled ventilation through small openings and weakness in the building fabric has a significant effect on the energy demands of buildings due to the leakage of conditioned air as well as infiltration of outdoor air that has not been conditioned to the appropriate temperature for thermal comfort. Conditioned air that is lost to the environment through leaks in the building fabric requires that additional air is introduced into the building which also requires conditioning, also air that infiltrates into the building can create discomfort for occupants, requiring further energy to bring the temperature back to a comfortable range. This places a greater demand on the HVAC systems, consuming more energy than would be required if the leaks were not present.

Increasing the airtightness of buildings to prevent the infiltration of unwanted air through the building fabric means that air that has been conditioned for thermal comfort remains within the building. Equally, air lost through the building fabric is reduced. Improving the insulation of the building fabric prevents unwanted thermal transfer to the external environment. During winter months, high levels of insulation reduce thermal losses through the building fabric, maintaining higher indoor air temperatures without the need for additional energy input to raise the air temperature. During summer months, insulation prevents the warmer external air from heating the internal air, reducing the need for cooling.

Though increasing the airtightness of a building to prevent leakages and infiltration of air is a highly recommended practice, it can lead to potential health risks due to the build-up of harmful pollutants if the air is not routinely ventilated. The lack of infiltration of fresh outdoor air could potentially lead to the hourly air change rate falling below the recommended value, resulting in poor indoor air quality (IAQ). Prolonged exposure to poor air quality within a building can lead to occupants experiencing Sick Building Syndrome (SBS) which can have serious negative effects on the health of occupants [7]. Though prolonged exposure to poor IAQ can result in SBS, contact with air containing concentrations of pollutants above recommended values for a moderate amount of time can reduce mental performance and result in lower concentration levels [8].

With respect to these criteria of delivering recommended levels of ventilation at a temperature suitable for thermal comfort, mechanical ventilation systems are suitable solution to fulfil these conditions [9]. Mechanical ventilation systems enable buildings to be highly insulated and airtight whilst providing fresh outdoor supplied at the appropriate temperature for thermal comfort for occupants. Attempts to improve the technology are on-going with some success found in the integration of heat recovery technology to reduce the energy demand when altering the supply air temperature [10]. Heat recovery technology is used to transfer thermal energy from one airstream to another, thereby increasing the temperature of one airstream whilst lowering the temperature of the other. This process is highly beneficial when dealing with temperature regulation of indoor air as little to no energy is required to transfer the thermal energy depending on the system used. This is beneficial as less energy is required to condition the air to the appropriate temperature due to the lower temperature difference between the outdoor air temperature and required temperature.

Though developments have been made to mechanical ventilation systems to improve efficiency, substantial energy requirements are still necessary for operation. Fans and blowers are required to circulate air around ducts to be delivered to the required spaces and overcome the pressure loss caused by friction between the air and duct walls. This forced movement of air can be highly energy intensive [11]. The energy demand from mechanical ventilation systems can be significantly increased due to the addition of heat recovery technology if not properly incorporated into the design. As maximum contact time between the air and heat recovery device is necessary for maximum efficiency, the majority of heat recovery devices cause a high pressure drop in the airflow. To overcome a potential pressure drop and ensure that adequate ventilation is provided, additional fans and blowers are used. If the energy required by the additional fans and blowers to overcome the pressure is greater than the energy saved by the use of the heat recovery device then the total energy demand from the mechanical ventilation system increases despite the use of heat recovery technology [6].

Natural and passive ventilation are two, similar, ventilation strategies that are becoming more commonly used to ventilate buildings. These strategies are able to reduce the energy consumption of buildings with regards to the energy required for ventilation by using the effects of wind driven flow and air buoyancy [12]. Pressure differences created by obstructions in the path of wind flow force air through a building via a combination of driving and suction forces. Air buoyancy due is due to the varying density of air at different temperatures. Warm air is less dense and so rises; cooler air is denser and moves to replace the warm rising air at low heights. This can be taken advantage of for ventilation purposes, warm air is allowed to escape to the environment at ceiling or roof level and cool outdoor air enters the building at lower levels to replace the escaping air. By controlling this process, buildings can be ventilated with little energy requirement. Passive ventilation does not require input from user for operation for continued ventilation compared to natural ventilation. The most common passive ventilation system currently employed across a variety of buildings is the commercial wind tower system.

Commercial wind tower systems are passive ventilation systems, based on the traditional vernacular design of baud-geer. Baud-geer have been utilised by traditional cultures for centuries, predominantly in dry, arid climates, as a method of delivering ventilation and cooling to buildings and homes [13, 14]. Baud-geer wind towers have been redesigned with modern engineering processes and materials to improve efficiency and suitability for contemporary requirements.

Commercial wind towers are able to provide ventilation to buildings by the manipulation of pressure differences created by wind flow around a building and the wind tower. This is demonstrated in Figure 1. Assuming a wind direction 90° to a face of the wind tower, high positive pressure is generated on the windward face of the building and wind tower; this creates a driving force, forcing air through the wind tower into the building. Low or negative pressure is generated on the side and leeward faces due to the high velocity of air moving around these areas. The low or negative pressure areas create a suction force, drawing air out of the building through the wind tower as the differences in pressure are attempted to be equalised [15]. Wind driven flow is the primary mode of operation for a wind tower, however during instances when there is little wind movement; ventilation is still achieved by the buoyancy, or stack, effect.

The recent development of various aspects of commercial wind towers to improve efficiency and design has been well documented. Optimal designs of the louvers, shape and cross-dividers have all been investigated using a combination of experimental testing and computer simulation [16-18].

Despite the innovation of commercial wind towers used to freely ventilate buildings, open access to the outdoor environment means that air temperature that is needed to maintain thermal comfort within the building is difficult to contain due to the constant air change. Dampers can be used to prevent airflow passing through the wind tower and escaping to the environment. However, this negates the use of the wind tower ventilating the building and reducing pollutant build up. This relates to both the cooling and heating required to air temperature for thermal comfort. One solution is to integrate heat recovery technology into the ductwork of a commercial wind tower unit.

Some research has been conducted to determine the viability and effectiveness of integrating various cooling technologies into commercial wind tower design [19]. The primary focus of this work would be to make sure of this technology in hot climates where significant cooling is required. Other work focused on the mixing of indoor air with small volumes of outdoor to dilute pollutants and maintain clean air in buildings as a method of heat recovery in cool climates [20].

Heat recovery technology used to cool incoming air in hot climates and heat incoming air in cool climates has significant potential for energy reduction when integrated into a wind tower. By using a zero or low energy process to alter the temperature of incoming air to a level closer to thermal comfort for occupants, less energy is required from traditional systems for regulating thermal comfort. The operation of each of these systems can be seen in Figure 2.

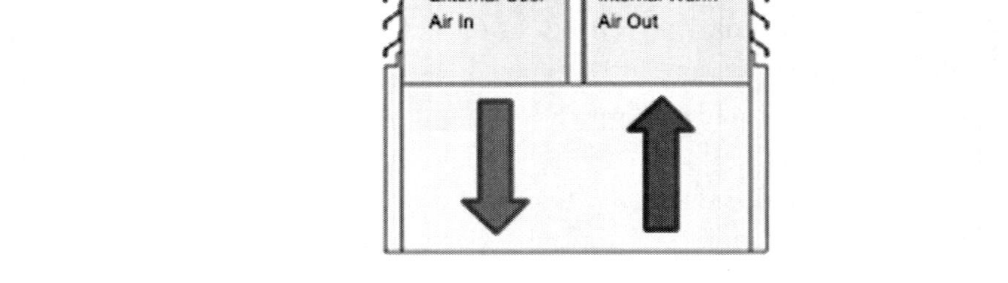

Figure 1. Schematic of Typical Commercial Wind Tower Operation.

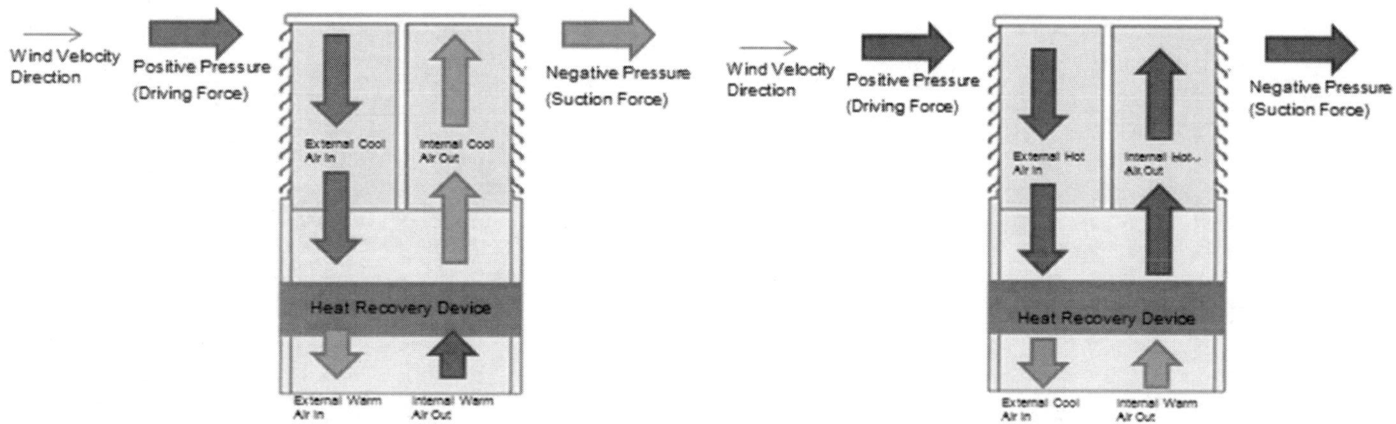

Figure 2. Schematic of wind tower with integrated heat recovery device a) heating mode, b) cooling mode.

Standard operation of wind towers in cool to mild climates is generally limited to summer seasons as the external air is too cold to be introduced into occupied spaces for the majority of the year. This restricts the operational time of the system to a relatively small window. The inverse of this situation relates to wind towers in warm to hot climates, operation is generally limited to the winter seasons when the external air is too warm to be introduced into occupied spaces for the majority of the year. Furthermore, research on the operation of wind towers in cooler climates shows that buildings which ventilate occupied spaces by natural means other than wind towers rarely achieve the required supply rates during summer with this value dropping significantly in winter [21].

Heat recovery technology provides a practical solution to reducing energy demand in mechanical ventilation systems by transferring heat energy from one airstream to another, thereby having a positive influence on the air temperature supplied to occupants in terms of thermal comfort. However, in order for the ventilation system to provide the required level of supply air to achieve adequate ventilation rate, additional fans and blowers are needed to overcome the pressure drop associated with the heat recovery technology. The additional energy demand of these fans and blowers can offset the energy saved by the heat recovery technology. Commercial wind towers offer a ventilation system which requires zero energy input to deliver the required ventilation rate but presently are only capable of operating when outdoor climate conditions are suitable to provide thermal comfort to occupants. Integrating heat recovery technology into commercial wind towers extends the window of operation for wind towers in all climates.

Provided that a building is constructed sufficiently airtight such that infiltration through the building fabric is negligible and the majority of supply and exhaust air passes through the wind tower system, the ventilation rate can be controlled and heat recovery technology would provide a method of controlled temperature regulation requiring very little energy input. The major obstacle to this conceptual design is the pressure drop caused by the airflow moving through/over the heat recovery device. Therefore it is pertinent to explore the range of heat recovery technology available which could be integrated into a wind tower and the advantages/disadvantages of each system. This gives designers and engineers a better understanding of which system would be most appropriate for integration.

WIND TOWER DESIGN

Traditional baud-geer have been used in Middle-Eastern countries with hot and arid countries to provide ventilation and cooling to buildings for many centuries. Natural and passive ventilation systems rely on wind driven airflow and the buoyancy effect to circulate air around a building, this has been thoroughly described by Linden [12]. Further to the benefits of cooling and ventilation that baud-geer provided, a constant supply of fresh outdoor air is beneficial to occupant health by reducing the concentration of airborne pollutants in indoor air. Examples of traditional baud-geer can be seen in Figure 3.

Improvements to traditional baud-geer have been investigated in an attempt to increase the effectiveness in providing airflow through the baud-geer along with modifications to improve the operation. Bahadori [13] enhanced the cooling potential of baud-geer by using evaporative cooling and used screens at openings to reduce dust infiltration. Using numerical and experimental analysis, Bouchahm et al. [14] made several modifications to baud-geer to improve the effectiveness of the system. The height of the baud-geer was increased and partitions within the shaft were introduced to increase the airflow rate. Wetted surfaces were provided to increase the cooling potential through the process of evaporative cooling.

Figure 3. Several ancient wind catchers with different configurations in the city of Yazd [30].

Though a traditional method of ventilation and cooling that requires zero energy input to provide comfortable conditions to buildings, the use of baud-geer has significantly reduced. Buildings that have existing baud-geer for ventilation are increasingly turning to mechanical air-conditioning systems to provide ventilation and thermal comfort [22]. This reflects a conscience effort for consumers to shift away from traditional technology to mechanical processes to fulfil the same needs. This is mirrored by the increased energy demand from HVAC systems and total global increased energy use. In response to this, designers and engineers are required to be innovative with solutions which are capable of meeting consumer needs, and more, but with lower energy demands.

The recent innovation of commercial wind towers is an example of bringing traditional solutions up to date with contemporary practices. Using modern engineering and design techniques applied to the design principles of baud-geer, commercial wind towers are able to fill the gap of zero-low energy ventilation solutions. Able to provide increased levels of airflow at a lower height, reduced dust and weather infiltration and operating at multiple wind directions, commercial wind towers advance the design of baud-geer significantly. A comprehensive review of wind tower development was conducted by Hughes et al. [15].

Wind Tower Development

Continuous development of wind towers to further increase efficiency, ease of use and operation time are vital design measures to maintain competitive status with mechanical ventilation systems. By ensuring that wind tower are as efficient as mechanical HVAC systems at delivering air to buildings but with zero energy requirements, adoption of the technology is likely to increase due to the benefits and cost savings to users. The most commonly used technique for development is Computational Fluid Dynamics (CFD). CFD allows designers and engineers to rapidly design, model and simulate new prototypes, components and configurations of wind towers compared to physical processes of manufacturing and experimental testing.

CFD simulations model the fluid flow around objects and the interactions, reactions and processes which occur, this is useful for the design of wind towers when using air as the fluid medium analysed. 3D CAD models can be created quickly and imported into CFD software for analysis. This process is significantly quicker and cheaper than experimental and sometimes is able to give real time results. However, it is important to note that experimental testing of physical models is still required to verify the CFD analysis due to the approximations made within the simulations. The benefit of CFD analysis ensures that only the optimal design identified by simulation is constructed for experimental testing, saving time and costs.

A significant amount of development of wind towers uses CFD analysis to improve the airflow rate through the wind tower. Optimisation of the louvers, shape of the wind tower and cross divider within the wind tower shaft all increase the airflow rate through the wind tower, thereby maximising efficiency. Control of the airflow is essential as the airflow rate through the wind tower increases, without control occupants could be exposed to uncomfortable draughts as air enters the building at ceiling height. Actuated dampers and purpose-designed grilles are used to prevent this.

The length and total number of louvers were optimised by Liu et al. [16] through the use of CFD analysis by evaluating the airflow rate through the wind tower and the uniformity of air distribution within the wind tower and building. The optimal number of louvers was found to be between 6-8, this was determined by the increase in airflow rate of 12.7% from an initial number of louvers tested as 4 louvers and 5 louvers. Increasing the number of louvers above this range only yielded an increase in airflow rate of 1.5%. Figure 4 visualises the airflow around the wind tower and the effect the number of louvers has on the airflow pattern and pathlines.

The angle of the louvers was investigated using CFD analysis and found to be similarly essential to the airflow rate through the wind tower as the number of louvers. Hughes and Ghani [17] incrementally varied the angle of the louvers by 5° between a range of 10-45° and analysed the effect on the airflow rate and pressure distribution within the adjoining building. From the results of the analysis it was found that the maximum internal airflow was measured when the louver angle was 35°. Above this angle, flow separation occurred around the louvers whereby airflow detaches from the surface of the louver, creating a wake behind the louver.

The plan shape of an object is a significant factor in determining the pressure difference that is created as wind moves over and around the object. This understanding of shape can be applied to the plan shape of wind towers to maximise the pressure difference and so maximise the airflow rate. Further to this, adapting the shape of a wind tower to accommodate the widest possible range of wind directions increases the operating conditions of the wind tower. In order to explore this design philosophy, Montazeri [18] tested a circular wind tower design with was configured with five different arrangements of cross dividers, this was done to determine the effect of increasing the number of internal sections on the airflow. The five designs were tested experimentally to validate numerical modelling previously performed. The designs were evaluated at a range of incident wind angles; the range was dependent on the number of sections the shaft was divided into so that at no time were more than two volumes exposed to the incident air angle. The results of the experimental testing can be seen in Table 1.

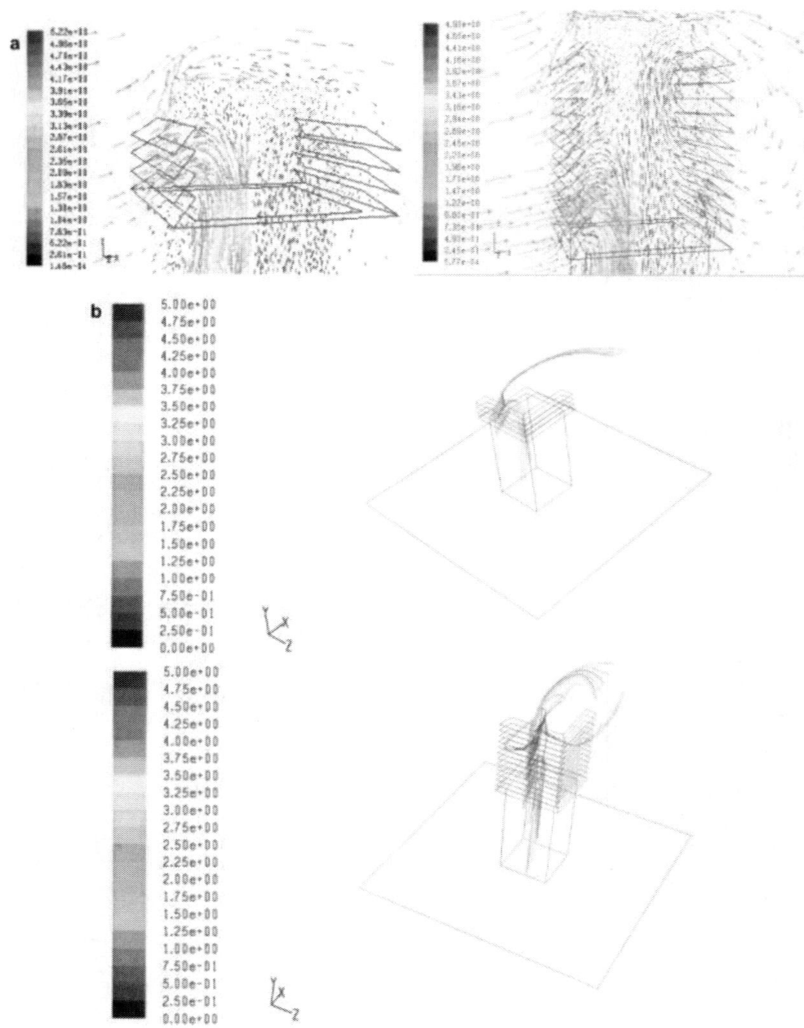

Figure 4. Airflow pattern and pathlines in and around the windcatcher (3-layer louvers and 12-layer louvers) (a) Airflow pattern, (b) Pathline [16].

Table 1. Number of Pitot and Static Tubes at the Bottom and at the Top Surfaces of the Models with the Related Angles [18]

Wind catcher model	Pitot tube		Static tubes		Wind Direction (deg.)	Openings
	Up	Down	Up	Down		
Two-sided	28	28	8	8	0-90	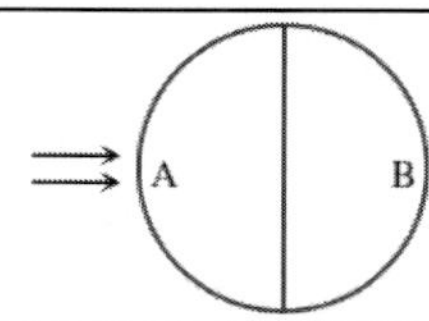
Three-sided	22	22	6	6	0-60	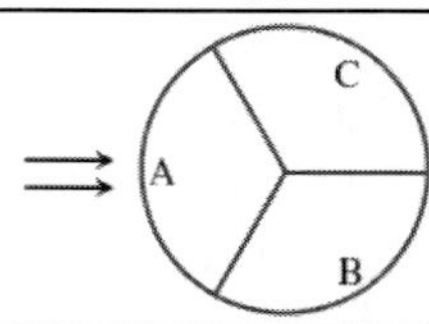
Four-sided	14	14	4	4	0-45	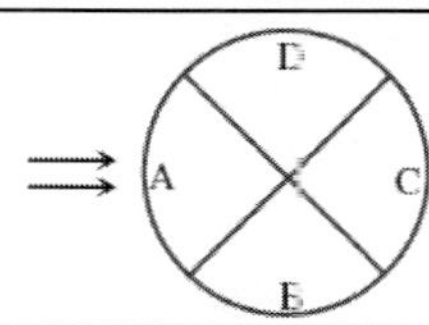
Six-sided	11	11	3	3	0-30	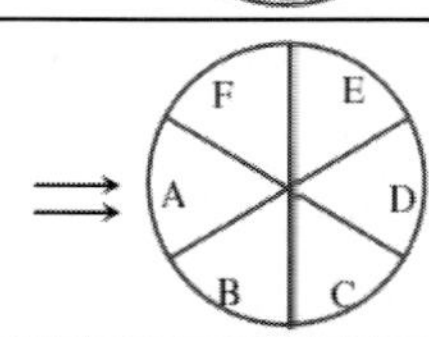
Twelve-sided	7	7	2	2	0-15	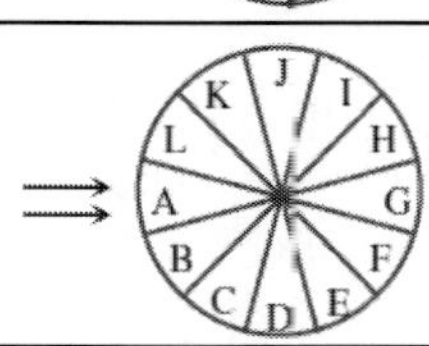

The results of the research did not yield an increase in airflow rate through the wind tower despite the modifications to the plan shape and number of cross dividers. The operation of a typical four-sided wind tower which uses two cross dividers, extending between each corner, to create four volumes for air to be supplied/extracted through often has two or more of the volumes used for exhaust air. Montazeri proposed that a wind tower using a greater number of cross dividers would result in more volumes used for supply air. Despite some logical basis behind the proposed designs, it was found that the four-sided wind tower was 13% more efficient than the circular design with same number of cross dividers. Further to this, at an incident angle of 0°, increasing the number of cross dividers reduced the airflow rate through the wind tower.

Despite commercial wind towers supplying adequate ventilation air based on guideline supply rates, enhancement of this airflow has been explored through the use of solar powered low energy fans installed in the shaft of a wind tower. Two separate research teams investigated the application of low energy fans in the shaft of a wind tower. Hughes and Ghani [23] used CFD to theoretically position a fan capable of inducing an additional pressure of 20Pa inside a wind tower at the top, middle and bottom of the shaft at an incident wind velocity of 1m/s. The results of the simulation showed that the low powered fan was capable of achieving the guideline ventilation rates at an incident wind velocity of 1m/s compared to the standard operation of a wind tower which cannot achieve adequate ventilation at 1m/s [24].

Priyadarsini et al. [25] used experimental testing to conduct a similar study involving an open-circuit boundary layer wind tunnel and scale models. Two wind tower models, one standard operation and a low-powered fan alternative, were fixed to a building model to determine the effectiveness of the fan in increasing the airflow through the wind tower. The data from the experiment showed that the low-powered fan alternative wind tower increased the airflow rate by 550% compared to the standard operation wind tower. The low-powered fan provided sufficient airflow without additional air velocity from the wind tunnel, highlighting the efficiency of the combined low-powered fan and wind tower system. It was suggested that the energy required to operate the low-powered fan could be supplied by a solar panel located on top of the wind tower, further enhancing the zero energy credentials of the system.

Management of the airflow rate into the building can be controlled through the use of dampers at the base of the shaft of commercial wind towers at times when the external climate conditions would cause significant discomfort to occupants due to draughts or high pollutant levels or used to accelerate airflow at time when external climate conditions would not provide enough ventilation air. Automated or manual control of the actuated damped would manage the angle of the dampers depending on the climate conditions and occupant demands.

The effect of dampers on the operation of commercial wind towers was conducted by two research teams using separate research methods. Using an experimental setup, Elmualim [26] concluded that the addition of dampers and egg crate grilles were capable of reducing airflow through a wind tower by 20-50% depending on the external wind velocity. The experimental setup can be seen in Figure 5. Hughes and Ghani [27] completed similar work using CFD simulation to determine the optimum louver angle for inducing airflow and pressure gradients. The dampers were arranged at 5° intervals for a range of 0-90°. Pressure and air velocity were calculated at a number of locations around the building model attached to the wind tower. It was found that as the angle of the dampers increased, the pressure drop across the dampers increased as did the air velocity. This is likely due to the acceleration of the air as it moves through the dampers. Using knowledge gained from this work, control of the angle of the dampers can be used to aid occupant comfort by reducing draughts at times of high external wind velocity or increase the airflow rate when external air velocity is low.

Wind Tower Testing

Wind tunnel testing is an analysis method which gives researchers the option to validate CFD simulations through the use of quantitative and qualitative data gathered from analysis of prototypes subject to the same conditions used in the simulations. This is useful and often a necessary step for researchers, designers and engineers in order to have confidence in the reliability of the CFD simulations. Due to the expensive nature of wind tunnel testing, both in terms of financial costs and time, CFD simulations are usually conducted until a solution deemed adequate is found before wind tunnel testing is conducted. However, with the recent developments of rapid 3D printing, creating prototypes for analysis is steadily becoming cheaper, quicker and more efficient. It is also noted by a number of researchers and designers that despite the best efforts of software development, CFD simulations can only be used as predictions compared to wind tunnel analysis and real world testing.

Wind tunnel experiments were completed to evaluate buoyancy driven ventilation through wind towers by Walker et al. [28]. A scale building model, designed from an office building with energy consumption and heat generation data available, was constructed and mounted within the wind tunnel; Figure 6 shows the experimental setup. Temperature and air velocity were measured at various points around the building model and wind towers. The experimental setup was modelled and simulated in CFD software. High levels of correlation were shown between data collected from the wind tunnel experiments and CFD simulations. This gave designers confidence in the CFD simulations to make alterations to the design and trust the results.

Figure 5. Experimental test set-up in the wind tunnel [26].

Figure 6. Picture of the scaled air model in the test chamber [28].

A major benefit of wind tunnel testing is the ability to visualise flow in real-time as flow develops using smoke visualisation and PIV techniques. Results from this analysis can easily be compared to contour plots and streamline patterns from CFD simulations. A number of studies have used wind tunnel testing for visualisation of air flow in one and two sided wind towers. Montazeri and Azizian [29] and Montazeri et al. [30] constructed wind tower models at 1:40 scale for testing in an atmospheric boundary layer open wind tunnel. The aim of the work was to determine the effect of incident wind angle on the performance of the wind towers. One sided wind towers are more commonly used in areas with a prevailing wind direction and require openings within the building such as windows or doors for the air to exhaust out of. Two sided wind towers provide more flexibility in terms of wind direction and building design as the two sides of the wind tower can be used as the inlet/outlet respectively. The data from both studies showed expected data where the maximum supply ventilation was achieved when the wind direction was perpendicular to the opening of each wind tower at a wind incident angle of 0°. The minimum supply ventilation was found when the wind incident angle was 90°, parallel to the opening of the wind towers. It was noted by Montazeri et al. [30] the likelihood of short circuiting was more common in the study of the two sided wind towers than in one sided wind towers, particularly at an incident wind angle of 60°. Short circuiting is the effect where air moves from the inlet to the outlet without entering the occupied space and providing ventilation.

Full scale, far-field testing is the final stage of development many technologies undergo in order to assess the application and functionality of a system in a real world scenario. Though wind tunnel technology and CFD analysis have developed to a sufficient degree to give strong indicators to system performance, the data collected from such analysis can only give reference to the performance due to the variability and random nature of real world testing. Far-field testing was completed for a wind tower system for a comparison with wind tunnel testing. Shea et al. [31] found that the ventilation air rate had a linear relationship with wind velocity. Analysis of the airflow through the four quadrants that made up the shaft of the wind tower due to the presence of two cross dividers provided useful insight to designers. The maximum ventilation rate was measure when the incident wind angle resulted in two quadrants acting as inlets for incoming air and two quadrants acting as outlets for exhaust air. When the incident wind angle resulted in one quadrant acting as an inlet for incoming air and the remaining three quadrats acting as outlets for exhaust air, the ventilation rate was measured at the minimum levels. However, though this was the minimum ventilation rate, the minimum guideline level was still achieved depending on the external wind velocity. A comparison between the real world analysis and CFD simulation showed that the CFD simulation consistently under-predicted the airflow rate. This highlights the importance of full scale far-field testing and over reliance on CFD analysis could lead to underestimation of performance of designs

Continued measurement of commercial wind tower performance used to ventilate operational, occupied buildings is necessary for development of wind tower systems. A number of natural ventilation systems, including commercial wind towers, were compared over the duration of a calendar year. Jones and Kirby [21] compared the ventilation rate of a variety of ventilation systems by measuring the CO_2 concentration in 16 different classrooms. The study found that commercial wind towers were a highly effective method of ventilation during summer months, and effectiveness increased when used in conjunction with open windows. However, during the winter months it was found that the classrooms that used wind towers as the primary method of ventilation suffered from high CO_2 concentration levels above the recommended guideline levels. The high concentration of CO_2 was due to the dampers angled at 0° from horizontal in order to prevent airflow through the wind tower, the dampers were controlled by temperature sensors within the classroom which determined that the internal air temperature was too low to introduce external air and would result in a loss of heat and thermal comfort.

The study recommended that CO_2 concentration monitoring sensors be used in place of air temperature sensors for controlling the damper angle. Though it was noted that this would likely cause a heat loss of 62W per occupant, this heat gain from occupant activity and electrical equipment would be enough to compensate for this. Furthermore, this conclusion gives credence to the integration of heat recovery devices to commercial wind towers for use in cool climates with cold winters as a method of introducing fresh outdoor air that is at a higher air temperature than the external climate. Hughes and Mak [32] used far-field experimental testing to study the wind pressure and buoyancy driven flows through a four-sided wind tower. The work examined the relationship between the two driving forces for the natural ventilation device. Figure 7 shows the commercial wind tower system mounted on top of a room in Sheffield Hallam University used for the full-scale experimental methods.

Commercial wind towers have been developed through modern, research led interest in optimising the design for maximum airflow rate through the wind tower. Using CFD simulation, wind tunnel testing and far-field testing to analyse prototype and design modifications, researchers and designers were able to increase the performance of commercial wind towers by concentrating on the components which affect airflow significantly. Far-field and real world testing has shown that commercial wind towers are capable of supplying the required ventilation rate to occupied spaces during summer months but better control of the dampers is required during winter in cooler climates in order to prevent discomfort to occupants due to heat loss. This opens up the possibility of integrating heat recovery devices into commercial wind towers as a method of conditioning the incoming air to an appropriate temperature thereby saving reducing energy demand. This can only be achieved however with the most appropriate selection of heat recovery technology.

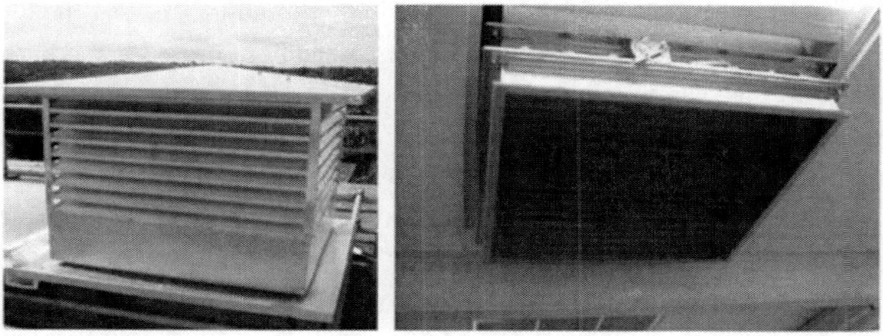

Figure 7. Far-field experimental setup [32].

Low Energy Heating and Cooling Technology

Heat recovery is a method which can be employed to reduce the heating and cooling demand currently place on mechanical HVAC systems. The aim is to alter the temperature of incoming air to a more suitable range for thermal comfort, though heat recovery technology may not be able to achieve thermal comfort completely but by conditioning the air closer to temperature required the additional energy input required by a mechanical HVAC system will be significantly reduced. Various different technology exists which are all able to achieve the same goal with varying processes and efficiency, therefore it is important to determine which technology provides the optimal design for integration into a commercial wind tower.

As exhaust air, which is likely to be at, or close to, the appropriate temperature for thermal comfort, is exhausted through the wind tower, it would pass through the heat recovery device. Here it would be used as a heat source or a heat sink depending on the climate conditions and needs of the occupants [33]. The thermal energy in the exhaust air can be transferred to the incoming fresh air, thereby raising the temperature and reducing the heating demand for climates that require substantial heating demand. For climates which require substantial cooling demand, the exhaust air can be used as a heat sink for the thermal energy in the warmer incoming fresh air, thereby reducing the cooling demand.

Heat recovery devices have been shown to significantly reduce the energy requirements of HVAC systems into which they are integrated by altering the air temperature in a low energy method [34, 35]. Due to the changing and evolving way in which buildings are being occupied and used, dominated by sedentary activity from occupants and electrical equipment, internal heat gains can be considerable purely due to the occupancy and operation of the building. The internal heat gains generated from occupant activity and electrical equipment could be utilised as a reliable heat source for a heat recovery system integrated into a commercial wind tower ventilation system.

Recovered heat can be divided into two separate classifications; sensible and latent heat. Sensible heat is the dry heat energy within the air. Latent heat is the energy recovered from the moisture of the air. Sensible only heat recovery devices generally have lower efficiency than total energy recovery devices which are capable of recovering sensible and latent heat [36-38].

Little work has previously been completed in highlighting the potential of heat recovery technology into a commercial wind tower ventilation system, instead focussing on the efficiency of heat recovery devices [33]. Mardiana-Idayu and Riffat [10] completed a thorough analysis of heat recovery technology and the application into both mechanical and passive ventilation systems. The type of ventilation system was the deciding factor in the appropriate selection for heat recovery technology; mechanical ventilation systems commonly employed heat pumps for heat recovery applications whereas heat pipes were effective for passive and natural ventilation due to the lack of moving parts and low pressure loss. It was noted that little work had been conducted in the area of combining of heat recovery and low carbon technologies such as evaporative cooling and desiccant dehumidification.

Evaporative Cooling

Evaporative cooling is one of the oldest techniques used for passive cooling in dry climates. This technique was used in old Middle Eastern cities and Persian Gulf states [39]. Since evaporative cooling is in some cases the only economical techniques for cooling large and semi-open areas such as atriums and central courtyards in hot climates, there has been renewed interest in the principles of such evaporative cooling techniques in recent years [40]. This has stemmed in numerous experimental works on evaporative cooling towers. Bahadori [13] proposed a new design to improve the cooling efficiency of natural ventilation wind towers. The design was capable of supplying air into the building at higher air speeds and with less dust. The analytical results indicated that the system was capable of reducing the temperature by up to 15°C and increasing the relative humidity by 50%.

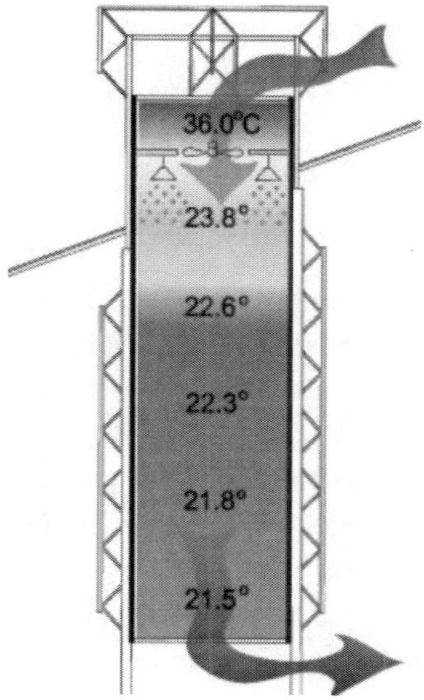

Figure 8. A wind tower with evaporative cooling spray on top [41].

Ford et al. [42] experimentally investigate the performance of a downdraught tower with evaporative cooling. The system was used to service a number of laboratories and office spaces within the building. Results from the tests indicated that indoor temperatures were 10-15 °C lower compared to the outdoor air temperature. The annual electrical energy consumption of the building was 64% lower compared to an equivalent mechanically cooled building, without affecting the comfort conditions of the occupants.

Bowman et al. [43] assessed the applicability of a downdraught tower with evaporative cooling for reducing energy consumption in hot dry climates. The assessment methodology employed dynamic thermal simulation programs for thermal analysis, and CFD modelling for airflow prediction. Preliminary results from CFD benchmark trials were presented. The evaporative cooling system was assumed to be capable of reducing the ambient air temperature by 70-80% of the wet-bulb temperature.

Montazeri et al. [44] assessed the performance of evaporative cooling provided by water spray system using the Lagrangian–Eulerian approach. Factors that influence the flow in the water systems such as velocity, pressure, humidity, droplet characteristics and continuous phase–droplet and droplet–droplet interactions were investigated. The evaluation was based on grid sensitivity technique and validation using wind-tunnel analysis. The results show that CFD simulation of evaporation was able to accurately predict the evaporation process, with local deviations from the wind tunnel measurements within 5-10%.

Heat Pipes

A heat pipe is a simple device of very high thermal conductivity with no moving parts that can transport large quantities of heat efficiently over large distances fundamentally at an invariable temperature without requiring any external electricity input. A heat pipe is essentially a conserved slender tube containing a wick structure lined on the inner surface and a small amount of fluid such as water at the saturated state. It is composed of three sections: the evaporator section at one end, where heat is absorbed and the fluid is vaporized; a condenser section at the other end, where the vapour is condensed and heat is rejected; and the adiabatic section in between, where the vapour and the liquid phases of the fluid flow in opposite directions through the core and the wick, respectively, to complete the cycle with no significant heat transfer between the fluid and the surrounding medium.

The operating pressure and the type of fluid inside the heat pipe depend largely on the operating temperature of the heat pipe. For example, if a heat pipe with water as a working fluid is designed to remove heat at 343K, the pressure inside the heat pipe must be maintained at 31.2kPa, which is the boiling pressure of water at this temperature. Though water is a suitable fluid to utilize in the moderate temperature range encountered in electronic equipment, various other fluids are used in the manufacturing of heat pipes to allow them to be used in cryogenic as well as high-temperature applications. Another characteristic while selecting the working fluid is the property of surface tension, which must be high in order to increase the capillary effect and being compatible with the wick substance, as well as being chemically stable, readily available, nontoxic and inexpensive. Figure 9 displays the basic working sections of a heat pipe.

Mathur [45] assessed the impact on overall energy consumption of treating ventilation air by retrofitting a heat pipe heat exchanger unit. Using the climatic conditions of Missouri, an in-depth investigation was conducted for the year round operation of the HVAC system equipped with the heat exchanger. The heat exchanger comprised of six rows of heat pipes in a horizontal orientation with an effectiveness of 60%. The study revealed that a heat pipe can be effectively used for increasing the efficiency of the existing HVAC systems.

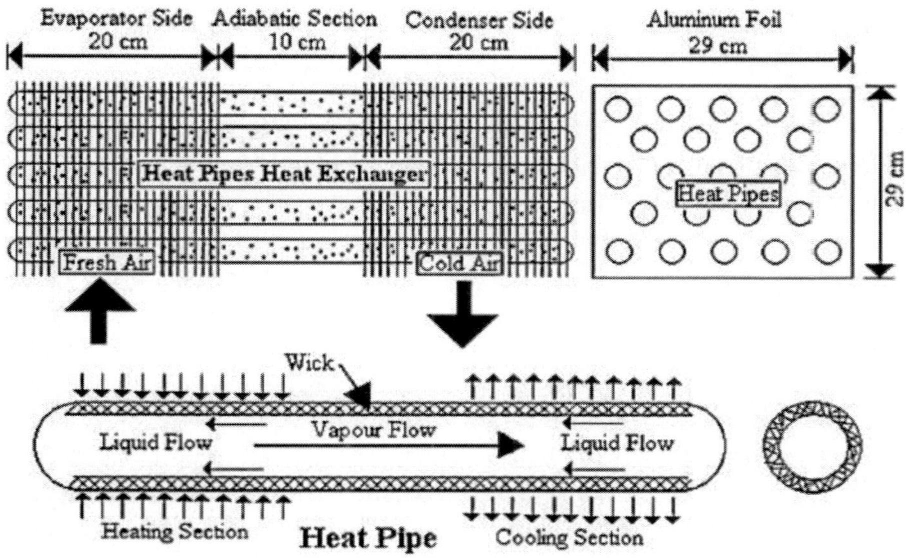

Figure 9. Heat pipe operation [45].

Wu et al. [46] evaluated the potential of heat pipes or thermosiphons as cold energy storage systems for cooling data centres. The emphasis of the study dealt with reducing electricity consumption of the facility. The study revealed that the system was capable of taking up to 60% of the total cooling load with a payback time of approximately 3 and a half years. In addition, with the reduction of external power consumption, the work revealed that up to 10.4 kilotons of carbon dioxide emissions can be reduced per year.

1:1 scale wind tunnel experimentation was carried out by Chaudhry and Hughes [47] to determine the responsive behaviour of heat pipes when exposed to varying hot ambient temperatures. The test replicated the hourly temperatures for June 21st, 2012 found in Doha, Qatar. The wind tunnel speed was kept constant at 2.3m/s to replicate a low Reynolds Number airstream (in the order of 10^3) typically found in natural ventilation systems. Data acquisition devices including K-type thermocouples and infrared thermal imaging systems were used in order to measure the dynamic upstream and downstream temperature. The mean heat transfer was recorded at 658W highlighting a heat pipe system effectiveness of 4.36% for the full duration of 24h. The study successfully classified the thermal performance of heat pipes operating under natural climatic conditions typically found in hot regions.

Rotary Thermal Wheels

Rotary thermal wheels are rotating wheels with a honeycomb or sinusoidal wave matrix which enables heat transfers from one airstream to another. The structure of the matrix enables air to move through the wheel whilst the contact time between the air and the matrix enables heat transfer to occur. The operation of a rotary thermal wheel acts similar to a temporary mass storage system, the temperature of the wheel matrix increases when subject to the airstream with higher temperature. As the wheel rotates, the section warmed by the airstream increases the temperature of the cooler airstream through convection [10]. The process repeats continuously as the wheel rotates. This is demonstrated in the schematic diagram in Figure 10.

Mathematical models of rotary thermal wheels have been developed to predict the performance of rotary thermal wheels based on variations in factors deemed to be necessary for system efficiency [49-53]. The majority of studies are validated by comparison with experimental testing or existing literature. The factors identified as relevant where the heat and mass transfer, effect of rotation speed, initial conditions of the two airstreams and the depth of matrix affecting air velocity and pressure loss.

The effect of the initial airflow velocity on the pressure loss across the rotary thermal wheel was considered by Hemzal [54] who carried out experimental testing to determine how flow was affected. The data collected from the experiment showed that the pressure loss across increased as the airflow velocity increased. High pressure loss across the rotary thermal wheel would render the integration of a rotary thermal wheel into a commercial wind tower for heat recovery unfeasible as the driving forces of the airflow would not sufficiently overcome the pressure loss to provide adequate ventilation.

Rotary thermal wheels can be designed to recover solely sensible heat or recover sensible and latent heat. Heat and total enthalpy wheels are similar in overall design, relying on the honeycomb or sinusoidal wave matrix to create contact time between the air and material for heat and/or mass transfer. However, the materials and processes used to construct the matrices differ for each type of wheel. Rotary thermal wheels are commonly constructed from aluminium or other materials with low specific heat capacity values in order to maximise the heat transfer. Enthalpy wheels can be constructed from similar materials for the matrix structure but then are coated with a desiccant material. Water is adsorbed onto the surface of desiccant materials; this becomes a physical transport of the moisture within the airflow between the two airstreams. The water adsorbed onto the surface of the desiccant material is desorbed when subject to a dry, hot airflow. It should be noted that enthalpy wheels can be constructed of other materials to increase sensible or latent heat transfer depending on the preferred operating conditions.

Adsorption and desorption have a significant effect on the air temperature of the two airstreams passing through the wheel. Adsorption increases the temperature of the airstream whilst desorption decreases the temperature. Both of these effects can have an impact on the energy consumption of HVAC systems. If an incoming hot and humid airstream is required to be cooled and dried for thermal comfort for occupants, the energy requirements will increase as the air temperature increases as the air is dried. Due to the increased contact time required for adsorption/desorption to occur, the optimum rotation speed for enthalpy wheels is significantly lower than heat wheels.

The efficiency of rotary thermal wheels is generally above 80% depending on the initial conditions of the airstreams the wheel is rotating through. Though efficiency is high when used as a standalone device in a mechanical ventilation system, rotary heat pumps are commonly used in conjunction to increase overall performance. Due to the efficient mass transfer process, enthalpy wheels are used in centralised air treatment units which supply air to whole buildings. Calay and Wang [55] found that this configuration of ventilation system reduced energy demand by 60% compared to conventional systems. Figure 11a compares the energy input required for a conventional air-conditioning system and a hybrid system as the outdoor air temperature increases. The hybrid system uses almost half the amount of energy compared to the air-conditioning system at the highest outdoor air temperature. Figure 11b compares the incoming air temperature of the two systems depending on the outdoor air temperature. The hybrid system was able to provide incoming air that was at a significantly lower air temperature than the air-conditioning system. The hybrid system reacts in a more linear, predictable manner as the outdoor air temperature increases compared to the air-conditioning system which experiences substantial increases as the outdoor air temperature increases. A separate study by Akbari and Oman [56] investigated the integration of a rotary thermal wheel into a ventilation system in a cool climate. The study found that energy savings up to 30-42kWh.m2 were possible.

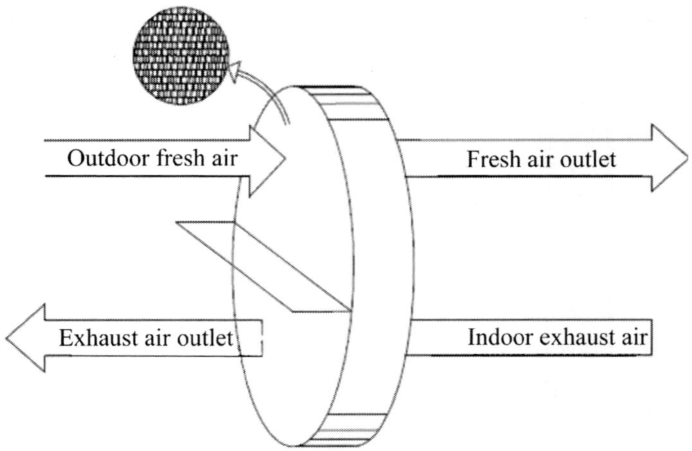

Figure 10. Working principle of wheel heat recovery unit [48].

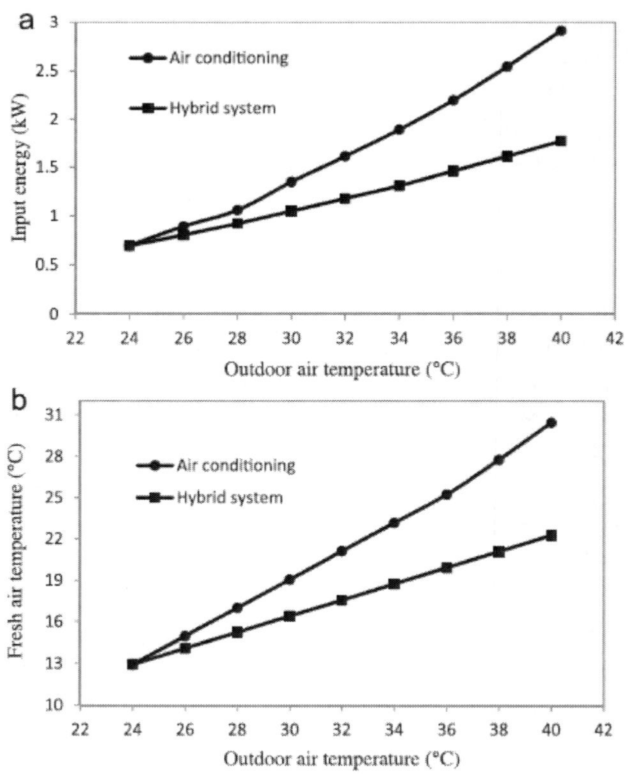

Figure 11. Effect of outdoor air temperature (a) on input energy required, and (b) on fresh air temperature [55].

Short-circuiting frequently occurs to airflow passing through a rotary thermal wheel. Air moving through a channel of the rotary wheel close to the partition between the two airstreams may not flow all the way through before being redirected back in the opposite direction. As a result of this characteristic, rotary thermal wheels are not suitable for situations when cross contamination of the airstreams is not acceptable. The use of seals and purge selectors can reduce the impact of the problem but do not isolate each airstream enough to completely remove cross contamination.

Fixed Plate Exchangers

Fixed plate heat exchangers consist of thin metallic plates stacked together with small gaps between each plate; two separate airstreams flow between the plates, commonly in adjacent directions but concurrent and counter-current flows can be accommodated. Lamb [57] stated that fixed plate heat exchangers operate best under counter-current flow conditions. Heat transfer occurs due to convection between the air and the plates. Fixed plate heat exchangers are highly efficient devices with regards to heat transfer. Values up to 90% sensible heat efficiency have been recorded for fixed plate heat exchangers due to the high heat transfer coefficients of the materials used, operational pressure and temperature range [58]. Fixed plate heat exchangers have been an area of research and development for 40 years [59].

Traditionally, the plates used for fixed plate heat exchangers are flat; Zhang [60] used corrugated plates for comparison. An example of the design used can be seen in Figure 12. The design decision behind the new design theorised that the troughs of the plate would increase heat and mass transfer due to the increased mixing and contact with the plate. Furthermore, thinner material could be used as a result of the increased strength of the triangular profile plates.

Condensation build-up on the walls of the plates is a common problem which affects the efficiency of fixed plate heat exchangers. If the air temperature falls below the dew point for the airstream, which is based on humidity and pressure, due to the heat transfer between the airstreams, condensation forms on the surfaces of the fixed plate heat exchanger. This can reduce the efficiency of the device and lead to further problems as reported by Fernandez-Seara et al. [61]. In climates where air temperature regularly falls below 0°C, frost and ice can develop from the condensate on the walls of the heat exchanger. This blocks the airflow through the heat exchanger reducing the efficiency. If the build-up of frost and ice becomes large enough, all flow through the heat exchanger could be blocked. A solution to this was developed by Kragh et al. [62], whereby the fixed plate heat exchanger continuously alternated between two modes of operation, preventing the build-up of frost and ice. The redesigned heat exchanger was able to maintain good operation and an efficiency of 82% even as external air temperature dropped below 0°C.

Membrane Fixed Plate Exchanger

Fixed plate heat exchangers are generally only capable of sensible heat transfer due to the materials used in construction of the device, focus on sensible and latent heat transfer through fixed plate heat exchangers has led researcher to use porous membranes for the plates, capable of both sensible and latent heat transfer. Figure 13 shows the schematic of a membrane fixed plate exchanger. The membrane is supported between two layers in order to provide structure to the plates; the supports do not interfere with heat or mass transport. Yaici et al. [63] used porous membranes as a material for total energy recovery devices. The results from this study showed that sensible and latent heat recovery is possible with membrane modelling across a fixed plate heat exchanger.

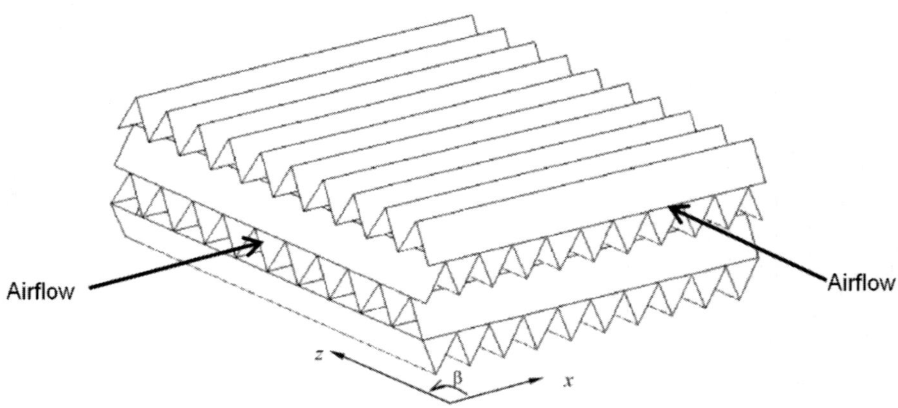

Figure 12. The flow channel geometry for corrugated fixed plate heat exchangers [60].

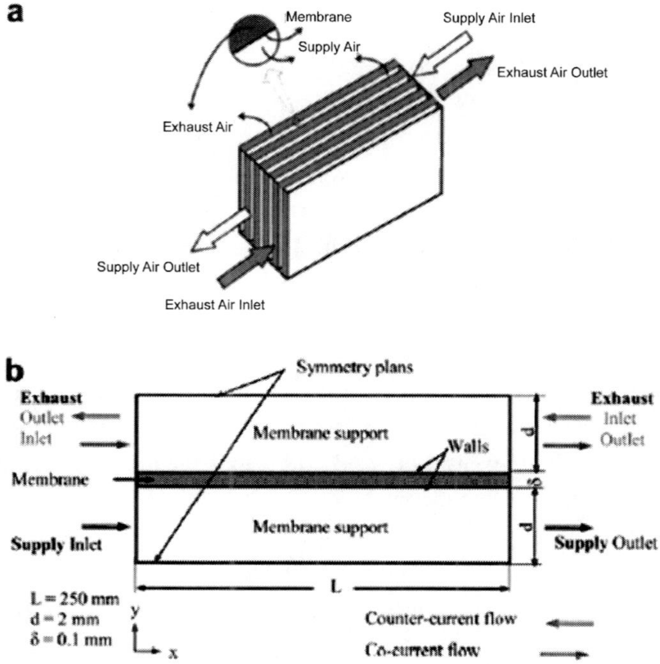

Figure 13. Schematic of a membrane-based energy recovery ventilator: (a) core in counter flow arrangement; (b) schematic of the physical model in a co-current and counter flow arrangements [63].

Supported liquid membranes (SLM) provide a large amount of potential for membrane fixed plate heat exchangers due to the heat and mass transfer properties [64]. SLM have a moisture diffusivity coefficient 3-4 order of magnitude higher than solid, porous membranes as shown by Zhang and Xiao [65], which enables greater mass and heat transfer. A significant amount of research, [66-68], explored the characteristics which affect the performance and efficiency of the fixed plate heat exchangers, those using SLM and those not. The plate thickness was found to be the most significant factor in heat exchanger efficiency though channel height between plates was found to be equally important. The efficiency of SLM fixed plate heat exchangers was calculated as high as enthalpy rotary wheels but without problems associated with frost and condensation due to developments.

Run-Around Systems

Run-around systems are a combination heat recovery system which makes use of multiple components to transfer heat from one airstream to another, potentially across a significant distance. A common setup for run-around systems uses two fixed plate heat exchangers connected by a closed loop containing fluid which is pumped around the loop. Heat is transferred through the fixed plate heat exchanger to the fluid. The fluid acts as an intermediary heat storage mass which is pumped to the fixed plate heat exchanger where the heat is transferred again [69], an example of a system can be seen in Figure 14. Though a run-around system requires a pump to circulate fluid around the loop, the energy requirements are lower than those required from fans to circulate air around a similar loop. Furthermore, the low pressure loss of the fixed plate heat exchangers requires no additional fans or blowers for airflow to pass through the heat exchanger.

The pressure loss caused by run-around systems was tested using two separate air-to-liquid heat exchangers connected by a liquid loop powered by a pump. It was found that a low pressure drop of 0.74Pa was experienced as air flowed across the heat exchangers when the efficiency of this system was 75.6%. The frictional losses caused by the fluid flow in the loop were important in determining the size, and hence the energy requirement, of the pump [11]. By increasing the allowable pressure loss across the heat exchangers, Davidsson et al. [70] were able to increase the efficiency of the run-around system to 80% for a 1Pa pressure loss.

Run-around systems are best utilised for ventilation heat recovery during early planning stages of a new development [71], the large size and equipment necessary for a run-around system generally prevent retrofit of existing buildings. Integration during the planning and design stage of a new building utilising passive ventilation is the most efficient and economical method [72].

Application of Low Energy Heating and Cooling Technology in Passive Ventilation

Buildings which use natural or passive ventilation system to provide supply air to occupants do not require energy input for operational purposes. However, the heating and cooling capabilities of these systems are commonly limited. Commercial wind towers particularly do not have current provisions for integration of technology to heat or cool the incoming air which is essential in improving the thermal comfort of occupants [15]. A number of attempts have been conducted to integrate low energy heating and cooling technology, often in the forms of heat recovery technology, into commercial wind towers to improve ventilation and thermal performance. A review of these techniques follows.

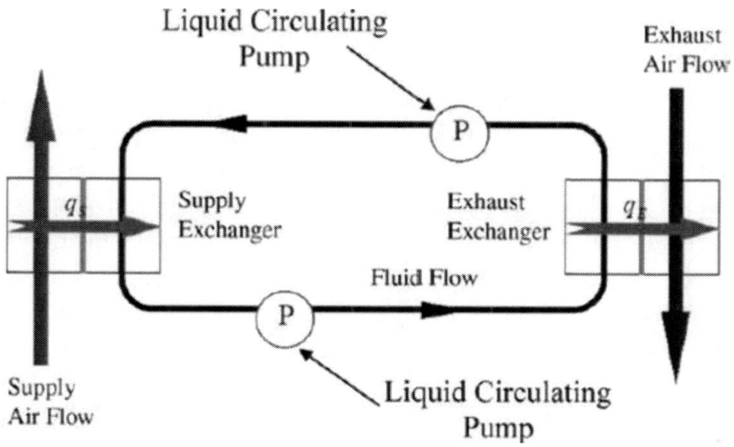

Figure 14. Schematic diagram of a run-around heat recovery system [69].

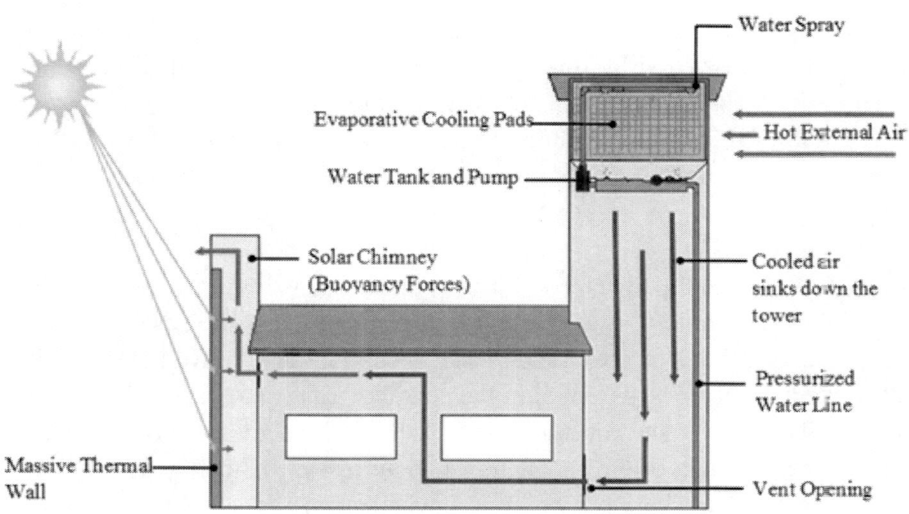

Figure 15. Concept design of a passive wind tower integrated with different cooling devices [15].

Low Energy Cooling

Figure 15 shows the concept design of a wind tower system integrated with cooling devices. Evaporative cooling pads sit at the top of the wind tower with a pump re-circulating water over them. Hot air is passed through these pads and cooled by water evaporation. Cool moist air is denser than ambient air and sinks down the tower and into the enclosed space. In order for the cool air to flow in, hot air must be released. A solar chimney is located directly opposite the wind tower to establish effective cross-flow ventilation inside the structure and exhaust the stored hot air using buoyancy-driven forces.

Wind towers equipped with wetted columns or wetted surfaces improve the ventilation and thermal performance of the passive devices and overcome the limitations of the conventional wind tower design. These towers can be employed in the hot arid regions and provide great saving in the electrical energy consumed for the summer cooling of buildings [73]. The evaporative cooling systems pre-cool the external air before admitting it into the structure. The cooled air becomes denser than the ambient air and sinks down the tower. Hence, air loss through other tower openings will be reduced. Wind towers incorporating wetted columns are equipped with cloth curtains or clay conduits, spaced 5-10 cm from each other and hanging vertically inside the column. The curtains are wetted by spraying drops of water through a nozzle system at the top of the tower. Wind towers with wetted surfaces are equipped with evaporative cooling pads at the entrance of the wind tower. Similarly the cooling pads are wetted by spraying water on top of the device. The cooling system is particularly suitable in arid regions with good winds.

Badran [73] also investigated the performance of an evaporative cooling wind tower system but chose to measure the air flow rates and internal temperature for a multi-directional tower. A mathematical model was developed to analyse the condition of air passing through the evaporative cooling column of the tower for different external conditions. Similarly, clay conduits were installed inside the tower's channel to cool the passing airflow before inducing it inside the structure as shown in Figure 16. During the night, the ambient air coolness is stored in the conduits mass to let it function during the day. The results showed that a 0.57 x 0.57m evaporative cooling tower with a vertical height of 4m can generate an airflow of $0.3m^3/s$ and reduce the internal temperature by 11K which is equivalent to the capacity of a 1 ton refrigeration system. Therefore, the author suggested that reducing the height of the wind tower which generally reached up to 15m can decrease the construction cost without having a noticeable decrease in performance. It was evident that the four sided wind tower was capable of reducing the indoor temperature and supplying air into the building at much higher air flow rates than the one-sided wind tower proposed by Bouchahm [14].

Correia Da Silva [74] evaluated the level of performance of passive evaporative cooling systems integrated in an auditorium building. A theoretical model was developed to predict the temperature and relative humidity of the air entering and leaving the structure through the cooling towers. The model was also used to demonstrate the influence of the physical parameters of the cooling device on the thermal environment within the building. The results showed that the performance of the passive cooling system was mainly dependent upon the evaporative cooling efficiency and number of cooling degree hours. The author concluded that the use of passive cooling methods in the summer is a suitable means of improving the internal thermal comfort and reducing the usage of conventional air-conditioning.

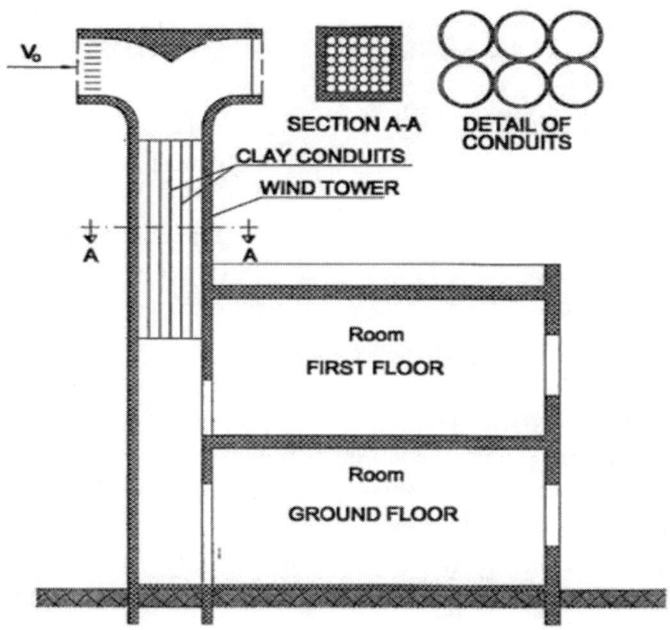

Figure 16. Wind tower integrated with wetted columns or clay conduits [73].

Bahadori et al. [13] evaluated the thermal performance of two novel designs of cooling tower systems using experimental testing. The two designs consisted of one with wetted columns, equipped with cloth curtains suspended in the tower and one with wetted surfaces was equipped with evaporative cooling pads at the entrance (Figure 17). The results established that the tower with a wetted column was more effective during high wind conditions while the tower with wetted surfaces was more effective during low wind conditions. The work concluded that integrating a cooling device to the conventional wind tower system proved successful, with the air exiting the towers at a significantly lower temperature than the external air. However, a small reduction in the airflow movement was observed inside the cooling tower.

Kalantar [75] attempted to evaluate the ventilation and thermal performance of a Badgir wind tower in the hot and arid region of Yazd. The work developed a numerical CFD model to simulate and analyse the airflow pattern inside the wind tower in three-dimensional and steady state conditions. The study also presented a numerical technique to simulate the effect of integrating evaporative cooling systems to the wind tower's performance. The effect of several design parameters such as wind speed, temperature, humidity and density were also considered. The result yielded a good correlation between the numerical simulations and experimental data obtained from literature. It was found that the Badgir wind tower was able to reduce the air temperatures by 10 to 15°C at its optimum performance (Figure 18).

Bouchahm et al. [14] evaluated the ventilation and thermal performance of a one-sided wind tower system incorporated to a climatically adaptable house using experimental and theoretical methods of analysis. The purpose of this investigation was to assess the potential of the evaporative cooling devices integrated to the passive ventilation system. Clay conduits were mounted inside the shaft of the one-sided tower to improve the mass and heat transfer and a water pool was situated at the bottom of the device to increase the humidification process (Figure 19). The analytical model was validated against the experimental measurements and a good agreement between the results was observed. The results confirmed that the airflow induced by the 0.75 x 0.70 m tower had a direct effect on the reduction of internal temperature. It was found that by using small sized partitions (created by increasing the number of conduits) better efficiency was achieved than with a higher wetted column of the cooling tower. The cooling tower integrated with wetted interior surfaces was able to reduce the indoor air temperature by up to 17.6 K depending on the height, diameter of the conduit partitions and climatic conditions.

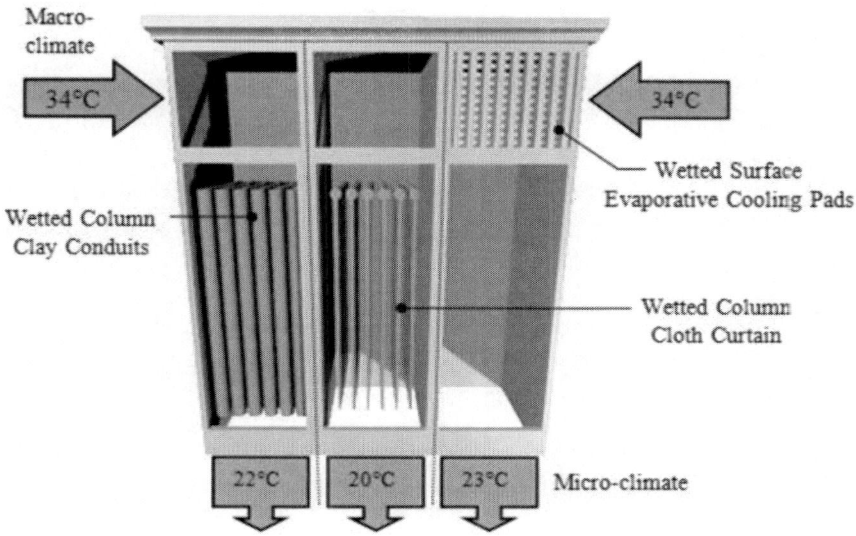

Figure 17. Thermal performance of a wind tower incorporating evaporative cooling devices [13].

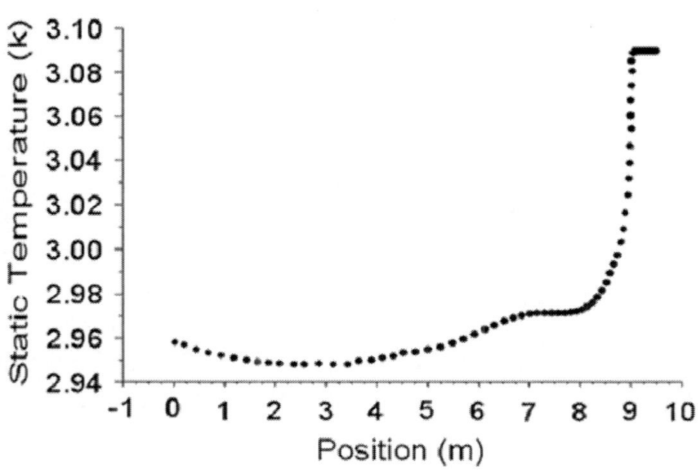

Figure 18. Variation of temperature of airflow from a 10 m high wind tower [75].

Bouchahm's research concluded that wind towers can provide a fresh supply of air and improve the thermal comfort of the inhabitants regardless of the extreme external conditions. The work demonstrated the significance of passive cooling towers and its potential as an alternative to the more prevalent mechanical ventilation systems.

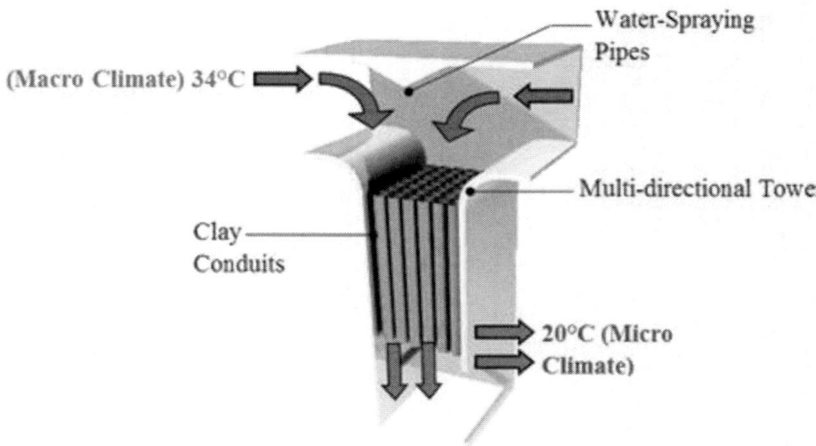

Figure 19. Wind tower integrated with wetted columns or clay conduits [14].

Furthermore, Saffari and Hosseinnia [39] used CFD modelling to investigate the thermal performance of new designs of wind towers under different structural parameters and external conditions. One novel wind tower design was equipped with wetted curtains suspended inside the column of the cooling device. These were formed as surfaces that injected droplets of water at extremely low speeds. A multi-phase CFD model, based on the Lagrangian–Eulerian approach, was used to study the effect of the diameter and temperature of the injected water droplets on the level of performance of the device. The numerical results showed that the 10 m high wetted columns were able to reduce the internal air temperature by 12° C and increase the relative humidity of the air by 22%. The study also revealed that a decrease in diameter of the injected water droplets resulted in a reduction in air temperature leaving the wetted columns. The small diameter droplets formed a larger evaporation surface area which led to better heat and mass transfer.

Low Energy Heating and Heat Recovery

The primarily role of natural and passive ventilation systems is to introduce fresh, outdoor air to buildings and remove pollutants. Because of this, operation of systems is generally limited to the cooling seasons in cool, temperature climates. Introducing outdoor air at low temperature could potentially cause thermal discomfort to occupants and may encourage the perception that natural and passive ventilation will increase heat loss and increase energy costs which are unfavourable to building operators. There are examples of natural ventilation in use during the heating season but this is generally restricted. However, by restricting the use of natural ventilation methods during winter months, the concentration of pollutants has been seen to rise above the accepted guideline levels and lead to poor mental performance and ill health [7]. To improve the year-round capabilities of natural ventilation systems to enable consistent use during cooler months, a retrofit heat recovery system is desirable [76].

Shao and Riffat [77] were the first to investigate the use of heat pipes in passive ventilation for heat recovery. They used CFD to evaluate the pressure loss and flow distribution in a natural ventilation system with heat pipes. The effect of the position of heat pipes and the temperature were also investigated. Figure 20 shows the room with a natural ventilation stack that was used for the simulation. The results showed that for a heat recovery efficiency of 50% and flow speed of 0.5 m/s, the pressure loss across the heat pipes was only 1 Pa. This showed that heat pipes would not cause significant reduction of the airflow and can be integrated with natural ventilation systems. Furthermore, the temperature of the heat pipe had a small effect on the flow loss performance of the proposed system.

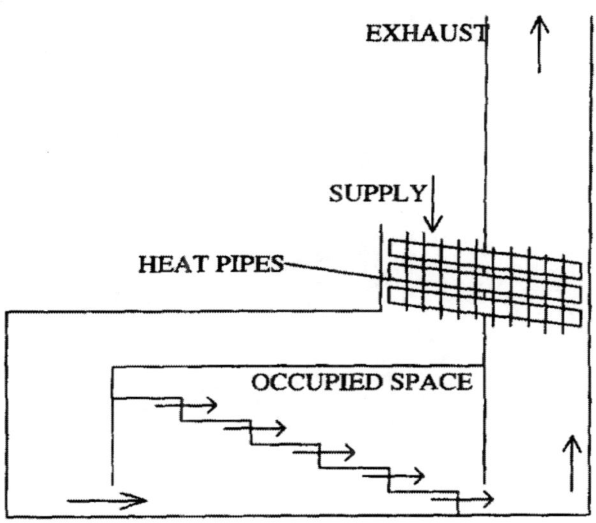

Figure 20. A room with stack ventilation and heat recovery [77].

Riffat and Gan [78] carried out an experimental study to investigate performance of heat pipe heat recovery devices integrated into a natural ventilation system. Additional numerical modelling was also conducted to determine the pressure loss characteristics of the heat pipes. The working fluid in the heat pipes was methanol with an operating range of -40°C to 100°C. Measurements of the heat pipe effectiveness were conducted in a two-floor test chamber as shown in Figure 21. The results showed that airspeed had significant effect on the performance of the proposed system. The effectiveness decreased with increasing airspeed. Higher effectiveness was achieved by using more banks of heat pipes although this increased the pressure loss. The work highlighted the increased in performance when fins were added to the pipes.

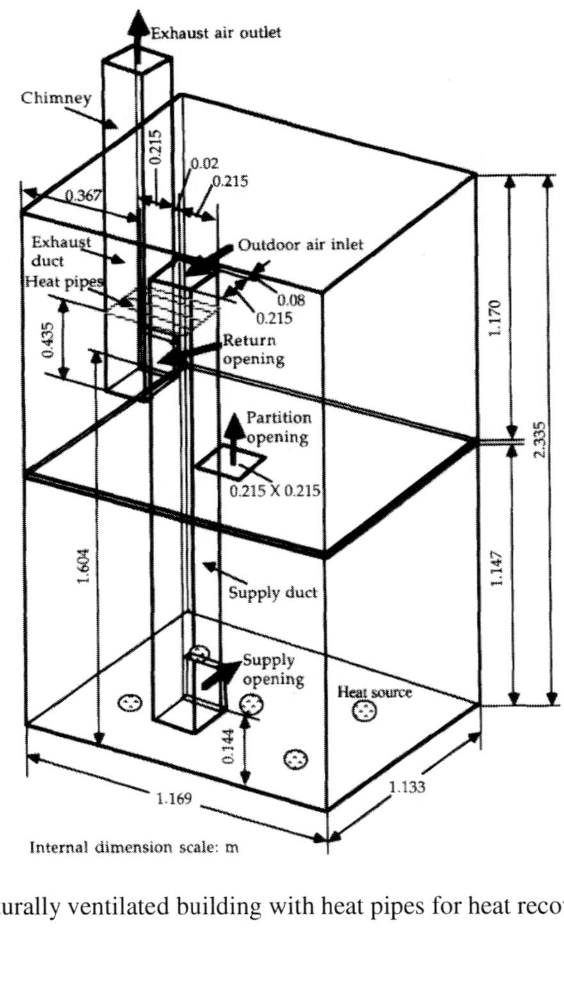

Figure 21. A naturally ventilated building with heat pipes for heat recovery [78].

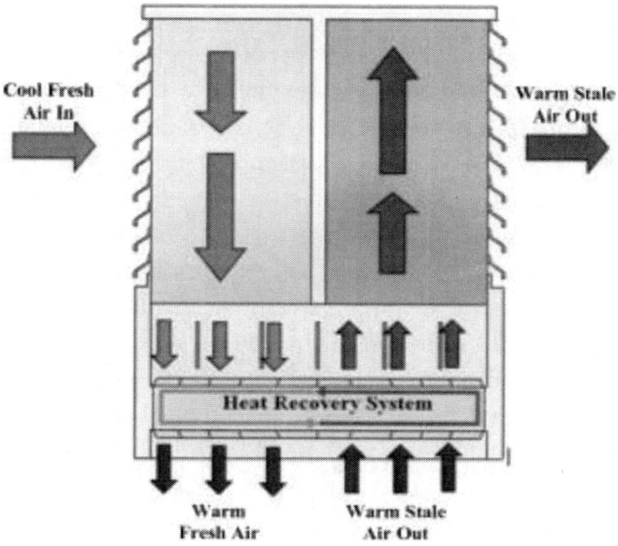

Figure 22. A wind tower integrated with a heat recovery wheel [76].

O' Connor et al. [76] proposed a new heat recovery method for a natural ventilation wind tower, a rotary thermal wheel. Using the properties of the rotary thermal wheel as a heat exchanger, the thermal energy in the internal exhaust air is recovered to the incoming air; this raises the incoming air temperature. By raising the temperature of the incoming air from the wind tower, adequate year round ventilation is maintained and during the heating season, energy demand for heating systems is reduced. CFD analysis was used to determine whether the airflow through a wind tower is significantly affected by the inclusion of a rotary thermal wheel at the base of the wind tower (Figure 22). CFD models were validated against experimental models tested in a closed loop wind tunnel for a wind tower concept that incorporated a rotary thermal wheel in the structure. The results showed that the wind tower with rotary thermal wheel was capable of meeting the guideline ventilation rates above an inlet air velocity of 3 m/s for a standard occupancy density of $1.8m^2$ per person. The potential to recover heat through a rotating thermal wheel could lead to further lower energy costs for heating systems due to the increased heat recovery capabilities.

Mardiana et al. [79] investigated the combination of a wind tower and a fixed plate heat exchanger under laboratory conditions. The experimental study was carried out using a chamber as shown in Figure 23. The results indicated that the heat recovery unit was capable of recovering part of the energy from the exhaust airflow. At air velocities ranged from 1.2 to 3.1m/s, heat recovery efficiency ranges from 70 to 50% for cold air condition and 69 to 49% for warm air condition. The maximum recovered energy of 772.9W was obtained at 5.2ºC temperature change of heat recovery unit. The work concluded that the proposed system can improve the energy efficiency and for long term it has the ability to be cost effective in energy saving.

Flaga-Maryanczyk et al. [80] conducted a study in Sweden which examined a passive ventilation system which integrated a run-around system which used a ground source heat pump as the heat source for incoming air. A schematic of the design is shown in Figure 24.

Measurements of air temperature and humidity were taken from the house which was used for analysis, along with relevant weather data used to build an accurate model. A 3d model was generated from the data and schematics of the house. CFD simulations were run to calculate the effectiveness of the run-around system and the relationship with the ground source heat pump. The ground source heat pump was capable of delivering 25% of the building's heating demand, working efficiently in conjunction with the run-around system.

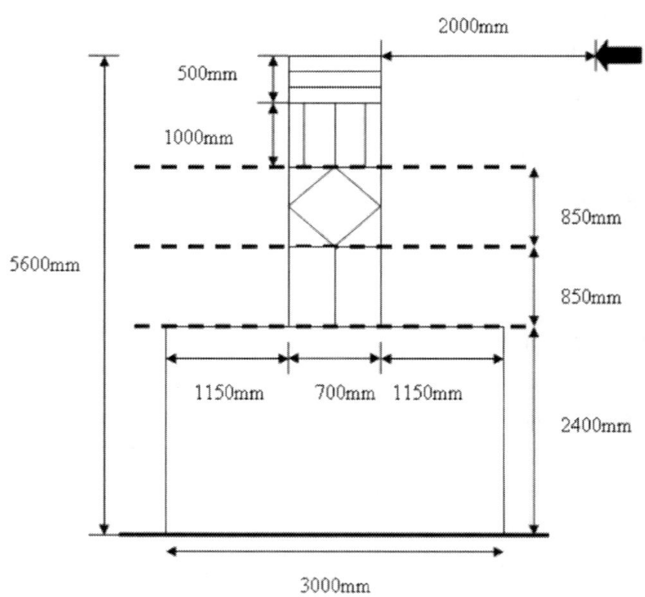

Figure 23. A wind tower integrated with a cross flow heat exchanger [79].

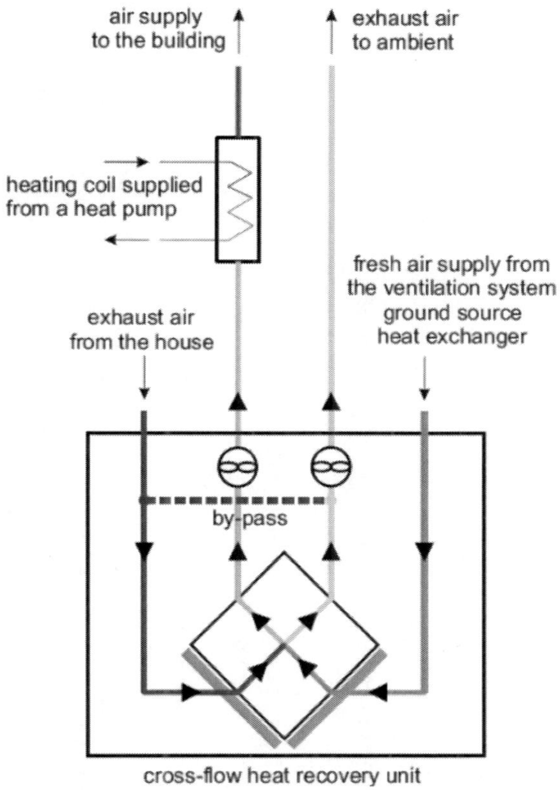

Figure 24. The schematic view of the ventilation system in the passive house [80].

CONCLUSION

Table 2 provides a summary of the low energy cooling/heating technologies reviewed in this work. The heat recovery devices are compared on a number of criteria, most notably advantages and disadvantages, efficiency range, pressure drop and humidity control.

Demand for low energy cooling and heating technology is becoming significantly more widespread for across a number of applications. The potential to significantly reduce the energy demand for cooling and heating from buildings by using low energy devices or recovering and transferring heat from one airstream to another. A review of these technologies has been undertaken to understand and appreciate the advantages possible.

Table 2. Summary and details of low energy cooling and heating devices

Type of Device	Advantages	Disadvantages	Performance Parameters	Efficiency %	Pressure Drop (Pa)	Humidity Control
Rotary thermal wheel	High efficiency Sensible and latent heat recovery Compact design Frost control available	Cross contamination possible Requires adjacent airstreams Mechanically driven, requiring energy input	Rotation speed Air velocity Wheel Porosity	80+	4-45	Yes
Fixed Plate	No moving parts hence high reliability High heat transfer coefficient No cross contamination Compact design Frost control possible Sensible and latent heat recovery	High pressure loss across exchanger Limited to two separate airstreams Condensation build up Frost build up in cold climates	Material type Operating pressure Temperature Flow arrangement	70-90	7-30	Yes
Heat pipes	No moving parts, high reliability No cross contamination Low pressure loss Compact design Heat recovery in two directions possible	Requires close airstreams Internal fluid should match local climate conditions	Fluid type Contact time Arrangement/ configuration Structure	80	1-5	No

Type of Device	Advantages	Disadvantages	Performance Parameters	Efficiency %	Pressure Drop (Pa)	Humidity Control
Run-around	Airstreams can be separate No cross contamination Low pressure loss Multiple sources of heat recovery	Multiple pumps required to move fluid Difficult to integrate into existing structures Low efficiency Cost	Exchanger type Fluid type Heat source	50-80	~1	No
Evaporative Cooling	High efficiency Temperature reduction up to 15 K	Reduced airflow High maintenance cost. Not effective in regions with high relative humidity	Spray mass flow rate Water droplet diameter Velocity Temperature Relative Humidity	N/A	~1	No

A combined system which uses low energy cooling or heating device and a commercial wind tower for ventilation could significantly reduce the energy demand from the operation and maintenance of buildings. Current mechanical HVAC systems account for up to 60% of the energy demand of a building, using low energy cooling/heating and passive ventilation could reduce this level by a significant amount or remove the need for mechanical HVAC systems completely, depending on the building and occupant demands.

Research has shown that integrating low energy cooling/heating devices into a commercial wind tower is a feasible solution to growing energy demand. Commercial wind towers are capable of supplying the required ventilation rates for incoming air and low energy cooling/heating devices can condition air to a suitable level for thermal comfort. The major obstacle to overcome is the pressure drop across the device, ensuring this value is kept as low as possible maintains adequate ventilation rates. Maintaining low pressure drop along with high efficiency are the primary concerns for researchers and designers. Alongside these concerns are humidity control, ease of integration and size.

At present, each of the heat recovery devices contribute to high pressure drop of airflow with the exception of evaporative cooling which does not cause a pressure drop due to the configuration of the device. Analysis has shown that heat pipes and run-around systems cause the lowest pressure drop when designed in the optimum arrangement. However, due to the large equipment and duct needs for run-around systems, these are unsuitable for the majority of passive ventilation systems, particularly commercial wind towers. CFD analysis and scale model wind tunnel testing show that rotary thermal wheels are capable of creating a low pressure drop in the airflow provided the matrix structure of the rotary thermal wheel was redesigned.

The technology with the most efficient heat transfer will lead to the greatest energy savings. Heat pipes and rotary thermal wheels provide the highest heat transfer between two airstreams. Moisture control of the airstreams is also important for thermal comfort. Rotary thermal wheels can be coated with desiccant material to improve the humidity control. Furthermore, technology which is capable of sensible and latent heat recovery has been shown to be the most effective. Evaporative cooling can reduce the air temperature by a notable amount and increase humidity which can be required in hot, arid climates.

Though rotary thermal wheels and heat pipes have been recognised as the most appropriate low energy heating method for integration into commercial wind towers, further research and development can increase the design and efficiency of each design for maximum heat transfer and minimum pressure drop. The structure of the matrix and depth of the wheel have significant bearing on the efficiency and pressure drop of rotary thermal wheels, by balancing these two characteristics of the rotary thermal wheel for passive ventilation and heat recovery, a rotary thermal wheel specifically adapted to passive ventilation can be designed. Heat pipes are capable of high heat transfer rates with moderately low pressure drop. Continued research exploring the optimum shape and arrangement of heat pipes would provide a suitable technology for integration into passive ventilation systems provided a suitable heat/cool sink could be guaranteed. Both systems are capable of transferring energy in both directions, summer cooling and winter heating would be possible within a single system providing year round ventilation and thermal comfort. Evaporative cooling only offers cooling to incoming supply air but is a suitable low energy technology for integration into commercial wind towers, used in conjunction with other low energy technology, could significantly reduce energy demand in buildings.

A review of current low energy cooling/heating technology has been conducted, identifying the key parameters which affect performance, the efficiency of each technology, the pressure drop cause by the technology and whether the system is capable of humidity control. The purpose of this review was to determine which technology, if any, would be suitable for integration into passive ventilation systems, specifically commercial wind towers.

Passive ventilation offers a zero-energy process to provide adequate ventilation into buildings. The savings made on electrical demand by replacing HVAC systems with passive ventilation has previously been noted. By coupling passive ventilation and low energy cooling/heating technology, an opportunity exists to significantly reduce energy requirements for ventilation, heating and cooling by using low energy technology to condition the incoming air to thermal comfort levels as desired by the occupants.

Of all the technologies reviewed, rotary wheels, heat pipes and evaporative cooling are the most feasible options for integrating into passive ventilation systems. High efficiency and the ability to transfer both sensible and latent heat are advantageous. The most important factor when considering what heat recovery technology to integrate into passive ventilation is pressure drop experienced across the heat exchanger. Due to the low velocity of air movement in passive ventilation systems, the pressure drop must be kept as low as possible in order to maintain adequate ventilation rates. Though at present rotary wheels and heat pipes do not fulfil this criteria, further research into the design of each technology could lead to advances which would enable integration into passive ventilation systems. Evaporative cooling does not cause the pressure drop noted by the other technology but is only capable of cooling the incoming air.

REFERENCES

[1] DECC. *Climate Change Act.* London, 2008.
[2] Tommerup, H. & Svendsen. S. (2006). Energy savings in Danish residential building stock. *Energy and Buildings*, *38*(6), 618-26.
[3] Liddament, M. W. & Orme. M. (1998). Energy and ventilation. *Applied Thermal Engineering*, *18*(11), 1101-1109.
[4] WBCSD. (2008). *Energy Efficiency in Buildings: Facts & Trends.* World Business Council for Sustainable Development.
[5] ORME, M. (2001). Estimates of the energy impact of ventilation and associated financial expenditures. *Energy and Buildings*, *33*(3), 199-205.
[6] Perez-Lombard, L., Ortiz, J. & Pout, C. (2008). A review on buildings energy consumption information. *Energy and Buildings*, *40*(3), 394-398.
[7] Clements-Croome, D. J., Awbi, H. B., Bako-Biro, N. Kochhar & M. (2008). Williams. Ventilation rates in schools. *Building and Environment*, *43*(3), 362-367.
[8] Mumovic, D., Palmer, J., Davies, M., Orme, M., Ridley, I., Oreszczyn, T., Judd, C., Critchlow, R., Medina, H. A., Pilmoor, G., Pearson, C. & Way, P. (2009). Winter indoor air quality, thermal comfort and acoustic performance of newly built secondary schools in England. *Building and Environment*, *44*(7), 1466-1477.
[9] Jones, B. M. & Kirby. R. (2010). The performance of natural ventilation windcatchers in schools - A comparison between prediction and measurement. *International Journal of Ventilation*, *9*(3), 273-286.
[10] Mardiana-Idayu, A. & Riffat, S. B. (2012). Review on heat recovery technologies for building applications. *Renewable and Sustainable Energy Reviews*, *16*(2), 1241-1255.
[11] Hviid, C. A. & Svendsen, S. (2011). Analytical and experimental analysis of a low-pressure heat exchanger suitable for passive ventilation. *Energy and Buildings*, *43*(2-3), 275-84.
[12] Linden, P. F. (1999). The fluid mechanics of natural ventilation. *Annual Review of Fluid Mechanics*, *31*, 201-238.
[13] Bahadori, M. N. (1985). An improved design of wind towers for natural ventilation and passive cooling. *Solar Energy*, *35*(2), 119-29.

[14] Bouchahm, Y., Bourbia, F. & Belhamri, A. (2011). Performance analysis and improvement of the use of wind tower in hot dry climate. *Renewable Energy*, *36*(3), 898-906.

[15] Hughes, B. R., Calautit, J. K. & Ghani, S. A. (2012). The development of commercial wind towers for natural ventilation: A review. *Applied Energy*, *92*, 606-627.

[16] Liu, S. C., Mak, C. M. & Niu, J. L. (2011). Numerical Evaluation of Louver Configuration and Ventilation Strategies for the Windcatcher System. *Building and Environment*, *46*(8), 1600-1616.

[17] Hughes, B. R. & Ghani, S. A. A. A. (2010). A numerical investigation into the effect of Windvent louvre external angle on passive stack ventilation performance. *Building and Environment*, *45*(4), 1025-1036.

[18] Montazeri, H. (2011). Experimental and numerical study on natural ventilation performance of various multi-opening wind catchers. *Building and Environment*, *46*(2), 370-378.

[19] Calautit, J. K., Hughes, B. R., Chaudhry, H. N. & Ghani, S. A. (2013). CFD analysis of a heat transfer device integrated wind tower system for hot and dry climate. *Applied Energy*, *112*(0), 576-591.

[20] Woods, A. W., Fitzgerald, S. & Livermore, S. (2009). A comparison of winter pre-heating requirements for natural displacement and natural mixing ventilation. *Energy and Buildings*, *41*(12), 1306-1312.

[21] Jones, B. M. & Kirby, R. (2012). Indoor air quality in U.K. school classrooms ventilated by natural ventilation windcatchers. *International Journal of Ventilation*, *10*(4), 323-338.

[22] Yaghoubi, M. A., Sabzevari, A. & Golneshan, A. A. (1991). Wind Towers - Measurement and Performance. *Solar Energy*, *47*(2), 97-106.

[23] Hughes, B. R. & Ghani, S. A. (2011). A numerical investigation into the feasibility of a passive-assisted natural ventilation stack device. *International Journal of Sustainable Energy*, *30*(4), 193-211.

[24] Calautit, J. K. & Hughes, B. R. (2014). Wind tunnel and CFD study of the natural ventilation performance of a commercial multi-directional wind tower. *Building and Environment*, *80*(0), 71-83.

[25] Priyadarsini, R., Cheong, K. W. & Wong, N. H. (2004). Enhancement of natural ventilation in high-rise residential buildings using stack system. *Energy and Buildings*, *36*(1), 61-71.

[26] Elmualim, A. A. (2006). Effect of damper and heat source on wind catcher natural ventilation performance. *Energy and Buildings*, *38*(8), 939-48.

[27] Hughes, B. R. & Ghani, S. A. A. A. (2009). A numerical investigation into the effect of windvent dampers on operating conditions. *Building and Environment*, *44*(2), 237-248.

[28] Walker, C., Tan, G. & Glicksman, L. (2011). Reduced-scale building model and numerical investigations to buoyancy-driven natural ventilation. *Energy and Buildings*, *43*(9), 2404-2413.

[29] Montazeri, H. & Azizian, R. (2008). Experimental study on natural ventilation performance of one-sided wind catcher. *Building and Environment*, *43*(12), 2193-2202.

[30] Montazeri, H., Montazeri, F., Azizian, R. & Mostafavi, S. (2010). Two-sided wind catcher performance evaluation using experimental, numerical and analytical modeling. *Renewable Energy*, *35*(7), 1424-35.

[31] Shea, A. D., Robertson, A. P., Levermore, G. J. & Rideout, N. M. (2010). Measurements of the performance of a wind-driven ventilation terminal. *Proceedings of the Institution of Civil Engineers-Structures and Buildings*, *163*(2), 129-136.

[32] Hughes B. R. & Mak, C. M. (2011). A study of wind and buoyancy driven flows through commercial wind towers, *Energy and Buildings*, *43*(7), 1784-1791.

[33] Sauer, Jr. H. J. & Howell, R. H. (1981). Promise and potential of air-to-air energy recovery systems. *International Journal of Refrigeration*, *4*(4), 182-194.

[34] Dodoo, A., Gustavsson, L. & Sathre, R. (2011). Primary energy implications of ventilation heat recovery in residential buildings. *Energy and Buildings*, *43*(7), 1566-1572.

[35] Ke, Z. & Yanming, K. (2009). Applicability of air-to-air heat recovery ventilators in China. *Applied Thermal Engineering*, *29*(5-6), 830-40.

[36] Gendebien, S., Bertagnolio, S. & Lemort, V. (2013). Investigation on a ventilation heat recovery exchanger: Modeling and experimental validation in dry and partially wet conditions. *Energy and Buildings*, *62*, 176-189.

[37] Rasouli, M., Simonson, C. J. & Besant, R. W. (2010). Applicability and optimum control strategy of energy recovery ventilators in different climatic conditions. *Energy and Buildings*, *42*(9), 1376-85.

[38] Zhang, L. Z. & Jiang, Y. (1999). Heat and mass transfer in a membrane-based energy recovery ventilator. *Journal of Membrane Science*, *163*(1), 29-38.

[39] Saffari, H. & Hosseinnia, S. M. (2009). Two-phase Euler-Lagrange CFD simulation of evaporative cooling in a Wind Tower, *Energy and Buildings*, *41*(9), 991-1000.

[40] Givoni, B. (1993). Semi-empirical model of a building with a passive evaporative cool tower. *Solar Energy*, *50*(5), 425–434.

[41] Ben-Gurion University of Negev, *Desert Architecture and Urban Planning*, [online] 2015 [Accessed 28/01/2015]. Available from: http://www.bgu.ac.il/CDAUP.

[42] Ford, B., Patel, N., Zaveri, P. & Hewitt, M. (1998). Cooling without air conditioning: The Torrent Research Centre, Ahmedabad, India, *Renewable Energy*, *15*(1-4), 177-182.

[43] Bowman, N., Lomas, K., Cook, M., Eppel, H., Ford, B., Hewitt, M., Cucinella, M., Francis, E., Rodriguez, E., Gonzalez, R., Alvarez, S., Galata, A., Lanarde, P. & Belarbi, R. (1997). Application of passive downdraught evaporative cooling (PDEC) to non-domestic buildings, *Renewable Energy*, *10*(2-3), 191-196.

[44] Montazeri, H., Blocken, B. & Hensen, J. L. M. (2015). Evaporative cooling by water spray systems: CFD simulation, experimental validation and sensitivity analysis, *Building and Environment*, *83*(1), 129-141.

[45] Mathur, G. D. (1996). Using heat-pipe heat exchangers for reducing high energy costs of treating ventilation air. *In: Energy conversion engineering conference (IECEC 96)*, 11-16 August 1996, Washington DC, USA. 1447–52.

[46] Wu, X. P., Mochizuki, M., Mashiko, K. & Nguyen, T. (2011). Cold energy storage systems using heat pipe technology for cooling data centres *Front Heat Pipes*, *2*, p. 013005.

[47] Chaudhry, H. N. & Hughes, B. R. (2014). Climate responsive behaviour of heat pipe technology for enhanced passive airside cooling, *Applied Energy*, *136*(1), 32-42.

[48] Zou, T., Ma, L. & Xue. Y. (2010). Experimental Study on the Performance of rotary heat recovery unit. *In: 2010 Asia-Pacific Power and Energy Engineering Conference (APPEEC 2010), 28-31 March 2010*, Piscataway, NJ, USA. IEEE, 2010, p.4.

[49] Nobrega, C. E. L. & Brum, N. C. L. (2009). Modeling and simulation of heat and enthalpy recovery wheels. *Energy*, *34*(12), 2063-2068

[50] Wu, Z., Melnik, R. V. N. & Borup, F. (2006). Model-based analysis and simulation of regenerative heat wheel. *Energy and Buildings*, *38*(5), 502-514.

[51] Yamaguchi, S. & Saito, K. (2013). Numerical and experimental performance analysis of rotary desiccant wheels. *International Journal of Heat and Mass Transfer*, *60*, 51-60.
[52] Zhang, X. J., Dai, Y. J. & Wang, R. Z. (2003). A simulation study of heat and mass transfer in a honeycombed rotary desiccant dehumidifier. *Applied Thermal Engineering*, *23*(8), 989-1003.
[53] Angrisani, G., Roselli, C. & Sasso, M. (2013). Effect of rotational speed on the performances of a desiccant wheel. *Applied Energy*, *104*(0), 268-275.
[54] Hemzal, K. (2006). Rotary heat exchanger efficiency influenced by air tightness. In: 17th Air-Conditioning and Ventilation Conference, Praha.
[55] Calay, R. K. & Wang, W. C. (2013). A hybrid energy efficient building ventilation system. *Applied Thermal Engineering*, *57*(1-2), 7-13.
[56] Akbari, K. & Oman. R. (2013). Impacts of heat recovery ventilators on energy savings and indoor radon in a swedish detached house. *WSEAS Transactions on ENVIRONMENT and DEVELOPMENT*, *9*(1).
[57] Lamb, B. R. (1982). Plate Heat Exchangers - A Low-Cost Route To Heat Recovery. *Journal of heat recovery systems*, *2*(3), 247-255.
[58] Nielsen, T. R., Rose, J. & Kragh. J. (2009). Dynamic model of counter flow air to air heat exchanger for comfort ventilation with condensation and frost formation. *Applied Thermal Engineering*, *29*(2-3), 462-8.
[59] Fehrm, M., Reiners, W. & Ungemach, M. (2002). Exhaust air heat recovery in buildings. *International Journal of Refrigeration*, *25*, 439-449.
[60] Zhang, L.-Z. (2005). Convective mass transport in cross-corrugated membrane exchangers. *Journal of Membrane Science*, *260*(1–2), 75-83.
[61] Fernandez-Seara, J., Diz, R., Uhia, F. J., Dopazo, A. & Ferro, J. M. (2011). Experimental analysis of an air-to-air heat recovery unit for balanced ventilation systems in residential buildings. *Energy Conversion and Management*, *52*, 635-640.
[62] Kragh, J., Rose, J., Nielsen, T. R. & Svendsen. S. (2007). New counter flow heat exchanger designed for ventilation systems in cold climates. *Energy & Buildings*, *39*(11), 1151-8.
[63] Yaici, W., Ghorab, M. & Entchev, E. (2013). Numerical analysis of heat and energy recovery ventilators performance based on CFD for detailed design. *Applied Thermal Engineering*, *51*(1-2), 770-80.
[64] Zhang, L.-Z. (2009). Heat and mass transfer in plate-fin enthalpy exchangers with different plate and fin materials. *International Journal of Heat and Mass Transfer*, *52*(11–12), 2704-2713.

[65] Zhang, L.-Z. & Xiao, F. (2008). Simultaneous heat and moisture transfer through a composite supported liquid membrane. *International Journal of Heat and Mass Transfer*, *51*(9–10), 2179-2189.

[66] Zhang, L.-Z., Liang, C.-H. & Pei. L.-X. (2008). Heat and moisture transfer in application scale parallel-plates enthalpy exchangers with novel membrane materials. *Journal of Membrane Science*, *325*(2), 672-682.

[67] Min, J. & Su, M. (2010). Performance analysis of a membrane-based energy recovery ventilator: Effects of membrane spacing and thickness on the ventilator performance. *Applied Thermal Engineering*, *30*(8-9), 991-997.

[68] Niu, J. L. & Zhang, L. Z. (2001). Membrane-based Enthalpy Exchanger: Material considerations and clarification of moisture resistance. *Journal of Membrane Science*, *189*(2), 179-191.

[69] Vali, A., Simonson, C. J., Besant, R. W. & Mahmood. G. (2009). Numerical model and effectiveness correlations for a run-around heat recovery system with combined counter and cross flow exchangers. *International Journal of Heat and Mass Transfer*, *52*(25-26), 5827-5840.

[70] Davidsson, H., Bernardo, R. & Hellstrom. B. (2013). Theoretical and experimental investigation of a heat exchanger suitable for a hybrid ventilation system. *Buildings*, *3*, 18-38.

[71] Emerson, W. H. (1983). Designing Run-Around Coil Systems. *Journal of heat recovery systems*, *3*(4), 305-309.

[72] Wang, J. C. Y. (1985). Practical Thermal Design Of Run-Around Air-To-Air Heat Recovery System. *Journal of heat recovery systems*, *5*(6), 493-501.

[73] Badran, A. A. (2003). Performance of cool towers under various climates in Jordan, *Energy and Buildings*, *35*(10), 1031-1035.

[74] Correia Da Silva, J. J. (2005). Passive downdraught evaporative cooling applied to an auditorium, *In: International Conference "Passive and Low Energy Cooling for the Built Environment"*, May 2005, Santorini, Greece, 555-560.

[75] Kalantar, V. (2009). Numerical simulation of cooling performance of wind tower (Baud-Geer) in hot and arid region, *Renewable Energy*, *34*(1), 246-254.

[76] O'Connor, D., Calautit, J. K. & Hughes, B. R. (2014). A study of passive ventilation integrated with heat recovery, *Energy and Buildings*, 82(1), 799-811,

[77] Shao, L. & Riffat, S. B. (1997). Flow loss caused by heat pipes in natural ventilation stacks, *Applied Thermal Engineering*, *17*(4), 393-399.

[78] Riffat, S. B. & Gan, G. (1998). Determination of effectiveness of heat-pipe heat recovery for naturally-ventilated buildings, *Applied Thermal Engineering*, *18*(3-4), 121-130.

[79] Mardiana, A., Riffat, S. B. & Worall, M. (2013). Integrated heat recovery system with wind-catcher for building applications: towards energy-efficient technologies A. Mendez-Vilas (Ed.), *Materials and Processes for Energy: Communicating Current Research and Technological Developments*, Formatex Research Center, Badajoz.

[80] Flaga-Maryanczyk, A., Schnotale, J., Radon, J. & Was, K. (2014). Experimental measurements and CFD simulation of a ground source heat exchanger operating at a cold climate for a passive house ventilation system. *Energy and Buildings*, *68*(PARTA), 562-570.

In: Natural Ventilation
Editor: Oscar T. Haynes
ISBN: 978-1-63482-414-9

Chapter 4

Improving Indoor Air Quality in School Buildings, Enhancing Comfort and Reducing Heating Energy Consumption through the Use of a Passive System

Jorge S. Carlos [•]
C-MADE, Centre of Materials and Building Technologies,
University of Beira Interior, Covilhã, Portugal

Abstract

School buildings constitute a great part of the non-residential building stock and it is particularly important to ensure a good indoor air quality therein. Natural ventilation is frequently the only ventilation system available in most of educational buildings. Therefore, ventilation rates are correlated with wind speed, wind direction, indoor-outdoor temperature difference and are closely related to building design. Unfortunately, this kind of ventilation is far from being reliable. The limited information available on the thermal comfort and indoor air quality, the absence of proper legislative measures and the lack of interest concerning ventilation efficiency may be some of the reasons for the non-resolution of this problem. School buildings face the same indoor air

[•] E-mail: jcarlos@ubi.pt.

quality problems as any other building. Although the health and productivity of pupils and teachers is strongly affected by the indoor environmental quality of their school, a recent survey conducted in Portugal, revealed poor indoor air quality in all the schools under study, even for recently constructed or refurbished buildings. Another survey conducted in Portugal also reported that thermal comfort in an elementary school is barely achieved. Thermal comfort models are usually based on adult subjects, being no assurance that these models can be applied to children. Surveys conducted in school children aged 7-11 in Hampshire and Southampton, UK on thermal comfort indicate that children in a classroom have a different thermal perception compared with adults, showing that this is a sensitive subject to deal with that requires further studies, to ensure a comfortable and healthy environment for educational purposes. Another study showed that installation of ventilation systems with heat recovery is often difficult or even impossible in existing school buildings. Therefore, natural ventilation plays a crucial role in promoting the comfort and health of occupants. It has the potential to significantly save cooling energy during cooling season, but it is costly in terms of heating energy consumption during winter, but can also be beneficial from an energy perspective when free heating is available. Different studies have shown that double ventilated windows can promote natural aeration of indoor environment with warmer air than it is found outdoors. Besides, this system can promote sustainability and can be use both on old and new buildings. The studies were based on objective evaluation by monitoring indoor conditions and subjective occupant evaluation using questionnaires. Simulations using analytical methods for different boundary conditions such as temperature and pressure differences across openings were also used to determine thermal comfort and energy savings. The potential of this ventilated heat recovery system is evaluated in terms of energy savings and thermal comfort. The present chapter deals with effects of wind and temperature difference between indoor and outdoor environments on the stack ventilation performance in an average elementary school in Portugal. The results for the cold moderate climate regions of Portugal and other coldest climates show that while indoor air quality can be easily maintained, the heating potential and thermal comfort strongly depends on the natural ventilation strategy.

INTRODUCTION

Healthy school environments are tremendously important to children's and teacher's health. This is a concern in many schools due in part to the age and poor indoor air quality (IAQ) conditions of a number of school buildings.

Maria do Carmo Freitas et al. [1] state that "Environmental asthma triggers commonly found in school buildings include respiratory viruses; cockroaches and other pests; mold resulting from excess moisture in the building; dander from animals in the classroom; and dander brought on the clothing from animals at home. Second-hand smoke and dust mites are other known environmental asthma triggers found in schools. Children with asthma may be affected by other pollutants from sources inside schools, such as unvented stoves or heaters and common products including chemicals, cleaning agents, perfumes, pesticides and sprays."

Clean air inside buildings has been the object of several studies within the scientific community, resulting in reports about buildings with air-related problems. Reported contaminants, especially carbon dioxide (CO_2) concentrations, volatile organic compounds and microbiological contaminants possibly increase the risks for allergen sensitivities, chronic irritation, cancer, asthma and 'sick building syndrome' symptoms [2]. This issue has also drawn the attention of the media [3]. There are reports relating that exposure in schools related to ventilation rates are similar to those in offices, and that children would be at least as sensitive as adults to these exposures [2]. However, this finding may not be consensus among researchers. Maria do Carmo Freitas et al. [1] state that children constitute a sensitive group with higher risk than adults and that IAQ problems in schools may be even more serious than in other categories of buildings. This is aggravates by crowded classrooms, poor construction and/or maintenance of school buildings. It can also reduce children's performance of schoolwork [4] and may be a cause absence from school [5].

Priscilla Nascimento Pegas et al. [6] performed measurements of outdoor and indoor pollution in three primary schools in Lisbon. These were considered representative of all the elementary-level educational institutions. All the classrooms depend only on the natural ventilation through the doors and windows existent, mostly closed during winter. They found that some indoor/outdoor ratios greater than unity and suggested that indoor sources and building conditions might have negative effects on air indoors. Poor air exchange rates, overcrowding and architectural finishes, are some sources of indoor contamination. Questionnaires among students to evaluate the association between the indoor air quality in schools and the prevalence of allergic and respiratory symptoms in adolescents were done in nine Porto schools [7]. They have concluded that CO_2, were associated with a greater respiratory symptomatology, which contributes to a lower indicator of indoor air quality. These problems are not reported only in Portugal. For instance, peer-reviewed literature indicates there is a clear indication that classroom ventilation is typically inadequate, and that average ventilation rate and ranges in the great majority of schools around the US and Europe are below of 5 l/s-person [2], minimum recommended by the ASHRAE Standard 62-1999 [8]. The extent of IAQ problems in schools, evidence shows that ventilation rates in schools often do not meet the minimum for health requirement, and this may be related to significant increases in symptoms among children and teachers in schools. Then adequate ventilation should be one of the major focuses of school design.

Teli et al. [9] state that current thermal comfort models are based on studies with adult subjects with no assurance however that these models apply to children. Later on [10] presented a study on children's thermal comfort in school classrooms. It is suggested that current adult-based comfort standards may not apply to school children. The adaptive comfort model was investigated by comparing the comfort temperature derived from European Standard EN 15251 [11]. It has shown that both models have limited suitability for predicting the thermal comfort conditions in schools as both did not accurately reflect the children's actual thermal sensation, especially outside the heating season [7]. Adjustments may be required to current comfort criteria in order to address the thermal perception of children [10].

Almeida and Freitas [12] state that the indoor environmental quality within a classroom is linked to the health, comfort and performance of students. A long term monitoring of several schools was carried on through winter, mid-season and summer conditions. Once most of the time temperature was within the comfort range of a retrofitted building mechanical ventilation systems were not being used. This had important consequences on the indoor air quality due to limitations in the window opening during the winter campaign. In a previous survey [13] the teachers reported that the windows were open during winter time when mild outdoor environment was verified. From what is said, Schools are the most suitable type of building for the application of energy efficiency and good indoor air quality measures, as state Theodosiou and Ordoumpozanis [14]. Lack of interest concerning energy efficiency and air quality problems in school buildings has been reported, since no action has been taken for years in order to improve the situation. This study is an exploratory analysis of the hypothesis that simultaneously optimizes energy and indoor air quality (IAQ) objectives can yield better results than existing natural ventilation control methods. There has been several different research conducted using numerical techniques to model airflow in a single room building [15, 16], being single-sided through open windows or cross ventilation. The chapter begins with a brief survey and discussion on interior airing and the consequent energy consumption, followed by the case study characteristics, the simulation methodology used for this study. Results are presented for different weather conditions to compare the effects of the temperature and wind on the airflow. Conclusions are withdrawn from the previous section.

AIRING AND ENERGY CONSUMPTION REDUCTION

As related before, ventilation of buildings plays a crucial role in promoting the comfort and health of its occupants. It is sometimes costly in terms of energy consumption, especially in cold winters. However, installing ventilation systems with heat recuperation in existing school buildings is often difficult or even impossible [17]. Ventilation air flow rates were simulated for different boundary conditions such as temperature and pressure differences across openings for different opening types, for three moderate climate locations in Germany, Italy and Turkey [18]. Different effective opening area of a tilted window was used during this study. A same three floor office building east or west orientated was situated in three different climates to evaluate the functionality of a controlled natural ventilated building under various external conditions. Carlos and Corvacho [19] presented a similar study were a ventilated double window was used, north or south orientated. Two different local sets of climatic data were chosen, a mild (Lisbon) and a cold winter (Oslo). Later on, Carlos [20] has studied the same ventilated double window on any façade and level and only one type of building. Four different weather conditions were chosen, from mild to cold winters, namely Norway, England, Germany and Portugal. Ventilation rates under various conditions depend mainly on the opening cross section, but also strongly on the location of the openings and their geometry. For instance, on a cross ventilation and chimney-supported buoyancy driven natural ventilation, result in high air exchange rates. Control over the openings is crucial in all cases to prevent overcooling of rooms and to provide enough fresh air in the heating season. The strength of the cross-ventilation will depend on the sum of the airflow resistances across the building. Still, its magnitude is largely determined by the smallest cross-section through which the airstream has to pass [21].

The opening of building facade has a strong influence on energy consumption, not only in terms of heat gain or loss but also in terms of ventilation. You et al. [22] have studied three type facade configurations with different areas or position changes. The opening position changes have a positive but small influence as around 2% of energy consumption. On the other hand, the area of opening has the most influence on a building's energy consumption. In addition, the window construction, the glass and frame materials should be selected carefully in the design stage. Stack ventilation is the only ventilation system available in most of the Portuguese primary school buildings. This is driven by the indoor and outdoor temperature difference and by the wind action on the building. In an indoor fixed temperature, the ventilation rate depends on the variation on the outside air temperature. As close are the indoor and outdoor temperature, the lower is the ventilation rate of the building. The direction and velocity of the wind may have an important effect on the ventilation rate, which may change up to 350% the ventilation rate [23]. Nevertheless, it is stated that stack ventilation is capable of ensuring appropriate IAQ inside buildings during winter. This could be closely related to building design, where ventilation shafts of proper design are favorable to augment the airing of buildings [24]. For instance, the air change rate varied between 8 and 15 with a solar chimney by inducing natural ventilation. Comparing the effect of opening the door or window with the inlet of solar chimney on the ventilation rate, using a single-room school turned south, the latter proved to be more efficient [25]. The natural ventilation of buildings may also be enhanced by the use of stacks. This is possible by connecting the floor with a relatively low heat load to a floor with a higher heat load through a common stack. The study done by Livermore and Woods [26] shows that the magnitude of the secondary ventilation within the lower floor increases with the ratio of the size of the openings on the lower to the upper floor and also the height of the stack. However, the secondary ventilation may lead to a reduction in the magnitude of the ventilation through the upper floor.

Most of the Portuguese school buildings are not equipped with suitable devices for natural and/or mechanical ventilation. The air changes are only due to air infiltration through the cracks and the frame of windows and doors, and at times, the occupants operate a control by means of windows and/or doors opening. In Portugal, a School Building Modernization Program was undertaken during the final of the last decade, where 106 public secondary schools were rehabilitated [27]. It included new integrated mechanical ventilation and air-conditioning systems regarding air renovation rates and comfort conditions. As a result a significant rise in energy consumption was achieved, being in some cases, up to 200%. In many cases the mechanical system was shut down with obvious disadvantages to the human health and comfort. As Khedari et al. [25] state, natural ventilation seems to be feasible and viable. A commitment must be obtained in order to maintain an acceptable IAQ from the health point of view, without compromising thermal comfort with minimum energy expenditure.

School building types of "Plano dos Centenários" constitute a major part of the elementary school building stock in Portugal (near 7000), built during the middle of the 20th century [28]. Within these buildings, no energy saving measures are applied to the operation of schools neither the control of IAQ. A representative number of this building stock could be studied, as, for instance, Dascalaki et al. did [29] for Hellenic schools. Monitoring indoor conditions and occupants behavior would give insight to pursue energy efficiency, thermal comfort and indoor air quality. The incorporation of passive systems could improve the performance of the building [28] as also could be much importance in terms of educational purposes. A first-step in defining possible intervention strategies to this kind of school buildings aiming energy efficiency and thermal comfort was already done [13]. The measures adapted to this kind of school building were defined according to the dimensions, technological and its architectural features. The analysis was performed taking into account the benefits achievable in terms of energy and money saving through a simple payback time analysis useful to identify priorities for action. Later on Santoli et al. [30] performed a similar analysis in public school buildings in Rome as well Krawczyk [31] did in Poland. This shows a growing interest in the issue.

Ventilation plays a crucial role in promoting health of building occupants being sometimes costly in terms of energy consumption, especially in winter time [4]. Heating up the ventilation air, as through an attic the surrounding rooms receive both pre-heated and sufficient fresh air, while the central atrium zone remaining warm [32]. Other strategies have also been studied. A recuperative and a regenerative ventilation unit were investigated by Manz et al. [33] through experiments and numerical simulations. The supply air is heated by a heat flow from the indoor air. It was found that for the regenerative and the recuperative ventilation unit, air change efficiencies of approximately 0.6 were determined at low levels of electric energy input. This was estimated by the temperature difference between in and out of the extract air. An opaque ventilated façade was studied in 12 locations in Spain in the winter [34]. This system allows the air to be entered through the lower opening of the external layer and introduced into the room through the upper opening. The external layer is made of steel plate that absorbs solar energy and transfers it to the ventilation system. The energy saving of this system depends strongly on the weather variables, such as the solar radiation on the façade, ambient temperature and wind speed. Energy saving increased as solar radiation and ambient air temperature increased. It was found that the sensible ventilation load of the heating system was positive for all the locations simulated in Spain during the winter season. Richman et al. [35] have studied a ventilated masonry façade wall to preheat the ventilation air. The air that is drawn from the outdoors, between the façade and sheathing, can be used to recover heat from the masonry. It acts like a solar collector. Besides the incident solar radiation, ambient conditions, air cavity velocity, and cavity dimensions are some of the factors that contribute to the amount of heat that can be recovered from the passing air. They claim that as a solar air collector can achieve as high as 33% daily solar efficiency and seasonal solar efficiencies of up to 27% and a seasonal energy savings of up to 48 kWh/m^2 for a standard residential dwelling. Carlos et al. [36] presented a variety of different technical solutions in order to compensate for thermal losses due to airing, where the incoming air is pre-heated passively before entering indoors. These systems include earth-to-air heat exchanger, enclosed spaces, solar collectors, different kinds of transparent facades. These systems can provide an effective way to preheat ventilation air at little to no extra cost. A double ventilated window that is included in the last category is going to be used in this chapter, showing through simulation that is possible to ventilate a classroom, reducing energy consumption and thermal discomfort.

CASE STUDY: A REPRESENTATIVE ELEMENTARY SCHOOL IN PORTUGAL

The restoration of buildings, especially public schools of national importance, is generally a complex endeavor as many parameters need to be considered such as preservation of its architectural identity. Complexity becomes even larger when issues of energy efficiency are needed to be taken into account. This may occur when new systems for heating, cooling and ventilation need to be installed, updating the building envelope or other issues related to reusing the building. One kind of building of particular interest is the elementary school built during the last century, in a program called "Plano dos Centenários", designed to fit all over the country to overcome the lack of school at that time. During the first half of the 20th century, the Portuguese government envisaged the construction of school buildings through all the national territory which were subjects of local materials and handcraft solutions. A typo of school project was designed as simple building [13]. The school was built with an innovative public concerns. It was design for natural ventilation and to take the most advantageous of solar radiation, both for heating and lighting. At that time, the concern of thermal comfort was only with heating, through a fireplace with a chimney in each classroom, thus it is not clear if the system ever functioned properly or efficiently. Lately, the firewood was not adequately controlled. The logging done by the parish, presented most of the time wood too green hindering combustion.

The kind of school building, from the “Plano dos Centenários” type were all based on the same plants and incorporated a series of sustainable local materials. They were built in a traditional way, mainly with stone walls and wooden floors and roofs, without any insulation material. However, some passive conditioning strategies were taken into consideration. The main façade orientation of these buildings were from East to South to take advantage from the solar radiation [13]. The large windows would allow solar gains and also daylight into the classrooms. The large windows of the main façade and a smaller one on the opposite façade of the school provide ample area for cross ventilation to occur within the classroom, when opened. The access to each classroom of the building is done through a common circulation with a staircase. Usually the front (main entrance on main facade) and rear (to the patio) doors are kept open to increase airflow through this section of the building, mostly to regulate temperature during the summer. Due to the stratification of the air with higher temperatures near the ceilings, the school is cooled using natural airflow through the buffer space, whereby air enters through the doors of the school and exits through the front balcony door on the first floor. Windows located in the classroom spaces on every level, can be opened to allow airflow through the classroom. These front large windows and smaller rear window are operated as the doors, only in summer time. During winter, all doors and windows are kept closed due to comfort problems.

The original building had no active thermal conditioning system, except for a small fireplace in each classroom connected to the chimney. Nowadays, in many school the traditional heating system was replaced by mechanical devices, providing heating only to the classrooms that keeps the interior environment at a set temperature. The circulation space is not provided with heating systems, as before, serving as a buffer space. Therefore, only the classrooms are defined as a thermal zone. In winter the heating system is complemented through the heat gains from the large glass panes, compensating for the heat losses resulting from the low exterior temperatures.

There is no active cooling system. The school building under study is found at Covilhã. The exact position of the building is latitude 40°17'18″N, longitude 7°30'10″ W, and an altitude of 631 m (Figure 1). The main façade is turned due Southeast (171 degrees from North). Its main facade combines large surfaces of windows (permeable to solar radiation and with low thermal inertia). The main geometric data of the classroom is shown in Table 1.

As regards natural ventilation, each classroom has three big windows on the main façade and a small usable window on the back façade for cross ventilation. For the present study, all the windows of the main facade are duplicated (Figure 2). Two parallel windows that fit each classroom fenestration. The outer window, which resembles the initial one, has an inlet at the bottom of the frame for the incoming fresh air. A controlled outlet is fitted on the original window (inner window) at the top of the frame to deliver warmer air in the classroom. In this way, the channel formed between the two window panes is used as an air path for incoming airflow. This double window system has a 3 mm single transparent glazed pane at the outer and the inner window, both made of wooden frame.

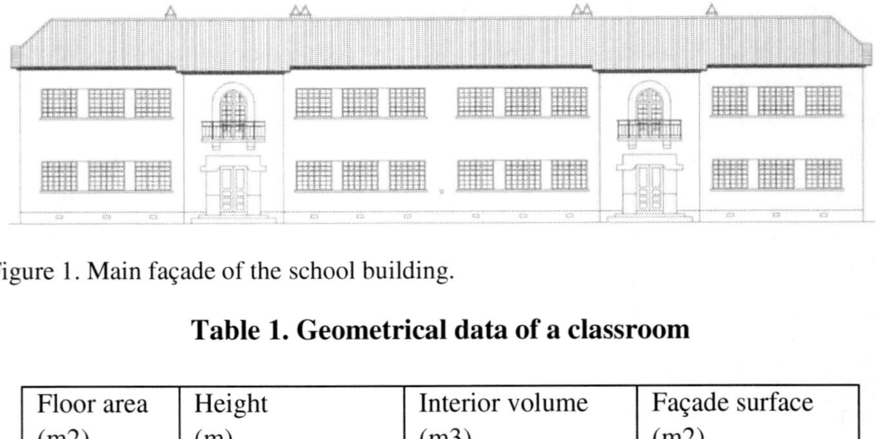

Figure 1. Main façade of the school building.

Table 1. Geometrical data of a classroom

Floor area (m2)	Height (m)	Interior volume (m3)	Façade surface (m2)
50	3.3	165	26.2
Windows surface (m2)	Window/façade (%)	Window/floor (%)	Glass/window (%)
8.3	31.7	16.6	74.0

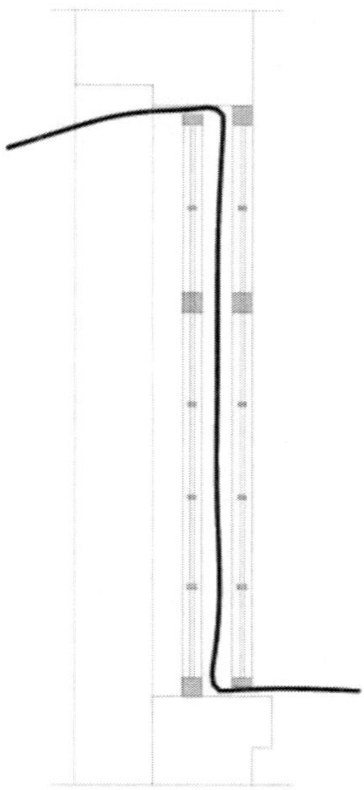

Figure 2. Ventilated double window – air flow pathway.

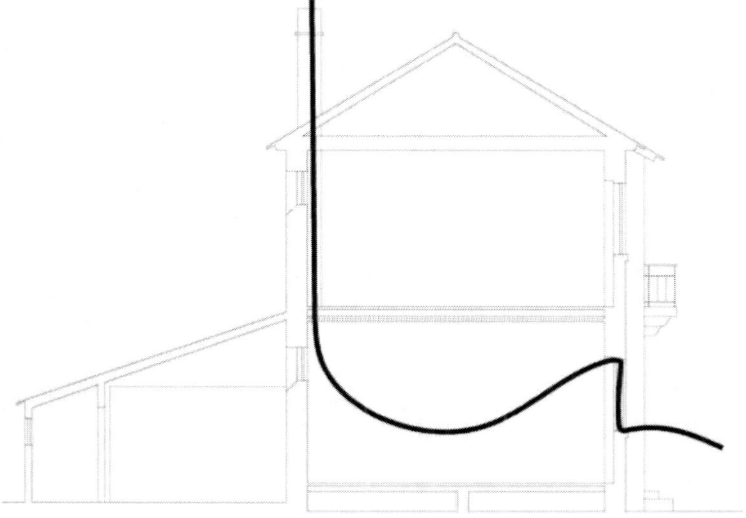

Figure 3. School building cross section – air flow pathway.

The average thickness of the air channel is of 7.9 cm. Glazing surface is about 73% of the surface of the whole window. Due to wind pressure and stack effect, the air rises between the two windows which is used as a pathway for incoming air. Having been pre-heated by the heat transferred from indoors through the inner window pane and by the solar radiation, the warmer air enters the building through the outlet at the top of the inner window. Once the fireplace is no longer used, it is now used for the exhausted air from the classroom, which goes up to the roof of the building (Figure 3).

DESCRIPTION OF THE SIMULATION METHODOLOGY

The methodology applied to examine the performance of a ventilated double window on the environmental conditions and the adaptation of this window system to the climate in which it is found through the energy and climate analysis. An energy models were built and validated to study this type of system [37]. The simulation provides data under specific climate conditions, quantifies its hourly evolution of air flow, useful heat that is recovered from the air stream and the air temperature at the outlet (air that enters the classroom).

The climate in Covilhã, Portugal is temperate, moderately cold in winter, being at the southern and Mediterranean climatic zone [38]. The climate data used for the simulation of the case study were taken from the Covilhã weather station, 3 km from the school building. Table 2 shows only the standard heating season weather data summary, i.e., the period from October until May, as defined by EN 832 [39]. The average monthly temperatures in Covilhã is between 6 °C in February and December and 15°C in May. The climate is such that heating is needed during the fall, winter and early spring. Cooling is needed during the late spring and summer months.

Given the typical occupant density of 25 per 50 m^2 (1 person for 2 m^2 of floor area) and a ceiling height of 3.3 m, the current ASHRAE standard 62-1999 [8] would require an air exchange rate of about 2.7 air changes per hour (ACH) for the classroom (5 l/s per person). As classrooms with higher number of windows/doors opened were associated with less polluted ambient optimized strategies could be employed to conventional natural ventilation methods [1]. When ventilation is no needed it could be closed down as also determining the minimum ventilation period required within a designated space [40]. In order to generate the results for the schedule occupancy of elementary school in Portugal, it is studied diurnal ventilation, so the openings (outlet of the window) are open only from 8 a.m. to 6 p.m. as also the heating system is.

Table 2. Covilhã weather data summary

	Jan	Feb	Mar	Apr	May	Oct	Nov	Dec
Global horizontal radiation (average daily – Wh/m2)	2107	2763	4205	6049	6139	2911	2567	2011
Dry bulb temperature (average monthly - ºC)	8	6	9	12	15	14	9	6
Wind direction (monthly mode – degrees)	30	50	30	20	20	200	30	30
Wind speed (average monthly – m/s)	2	2	2	2	2	2	3	3

In spite that that children respond in a different way to adults in their perception of overall comfort, more information is required on children's thermal comfort over a wider range of conditions and different ages [10]. Thus, the temperature of the classroom was set up to 22ºC to comply with the comfort standard [41].

The phenomenon of natural ventilation is strongly unsteady, depending on many factors. Being the geometry of the building defined, the air flow depends on the temperature difference between indoors and outdoors and also on the wind effect, as direction and speed. As said before, classrooms and circulation spaces are designed to enhance natural ventilation. The airflow it is of difficult to determine with accuracy [42]. The temperature and air velocity distributions within the classroom due to natural convection can be complex, especially when external factors such as wind patterns and direction influence the flow patterns within a building. Furthermore, air infiltration may occur through the building envelop (cracks, unsealed windows and doors and through the carpentry – wooden floor). It also occurs between different zones of the building, even when the communication door is closed (unsealed). Thus, a hypothetical sealed classroom is used to analyze the performance of this passive air heating system. After a global assessment of the performance during the entire heating season, four groups of five consecutive days are choosing to analyse the performance of the ventilated double window on this particular school. Each group is meant to represent a complete school week. The first two groups are chosen in function of the ambient temperature, meaning the lowest and highest temperature. The second two groups are chosen in function of the global solar radiation, meaning the lowest and highest solar irradiation.

ANALYSIS AND ASSESSMENT OF THE RESULTS

It was said before that natural ventilation is very unstable. This depends on the geometry of the building and on the local weather data. The latter cannot be changed. The building is an existing one with a geometry and orientation already defined. The windows are also defined. However, duplicating the window, vents are needed to be added to define the airflow pathway. The maximum possible vents were opened on each window without putting at risk the structure of each one. These vents are one of the geometric data that influences the air volume for airing the classroom. Once the geometry of both windows is defined and was assumed that the vents could not be enlarged, the focus should fall on another parameter to obtain the desired airflow rate (2.7 ACH). The chimney of the fireplace is another element of the stream pathway. The variation of its section will affect the amount of air volume for this cross flow ventilation. As natural ventilation is very unsteady, a series of previous simulations were run up. These aimed to obtain an average airflow rate during the heating season by adjusting the section of the chimney. In reality, it is very difficult to adjust a section of an existing chimney. Therefore, from the simulations, two identical chimneys should be added to the classroom to obtain the chosen ventilation. As the incoming air is pre-heated between the windows, the temperature of the delivered stream is higher than the ambient temperature. Figure 4 shows the daily mean global solar radiation (Solar), the mean hourly outdoor air temperature (Outdoor) and mean hourly temperature of the delivered air into the classroom (Outlet) for each month of the heating season.

The temperature of the delivered air is always higher than the outdoor air temperature. As lower is the outdoor temperature, the higher is the difference between the outlet and outdoor temperatures. The biggest difference is seen in February followed in December, months with the average lowest temperature. The relative difference between the outlet and outdoor temperatures can be found through the following equation:

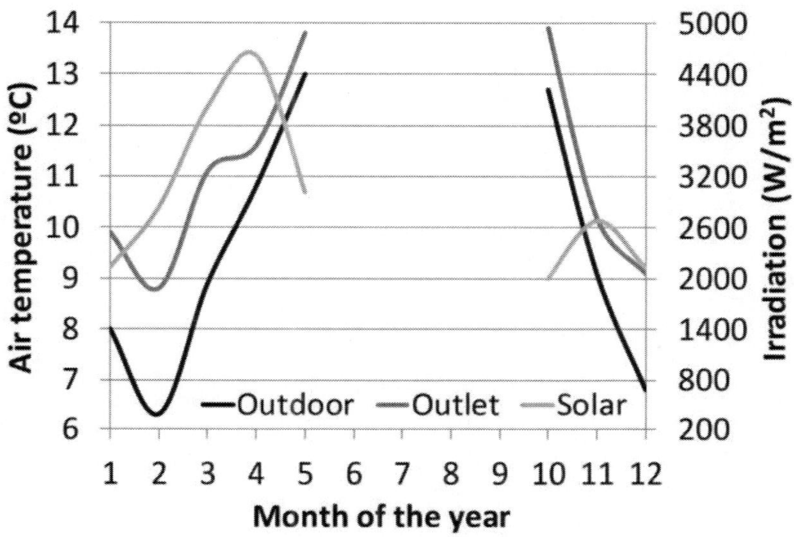

Figure 4. Heating season climate data and delivered airflow temperature.

$$\frac{\theta_1-\theta_2}{\theta_2}100\% \qquad (1)$$

where, θ_1 is the outlet air temperature (°C) and θ_2 is the outdoor air temperature (°C). During February the temperature rise is about 41% and in December of about 34%. The lowest temperature rise at the outlet is in May of about 6%. This is the month with the highest average outdoor temperature of 15°. The contribution to this temperature rise is not only from the incident solar radiation, but also from the thermal losses from inside through the inner window. This is a passive air heating system that acts as a heat reclaiming device. Part of the heat loss from inside through the window is removed from the passing stream that delivers it back to the classroom, acting as a heat recovery. Incident solar radiation warms the windows being part of that heat removed by the passing air before delivering it into the classroom, acting as a solar collector. In a previous study [43] it was found that solar collector function plays a small role in the pre-heating of the air, due to its transparency, which allows most of the solar radiation to enter directly to the indoor space. Therefore, the heat recovery is the most important function of the ventilated double window. That is why the relative difference between the outlet air temperature and outdoor air temperature is higher when the ambient temperature is lower.

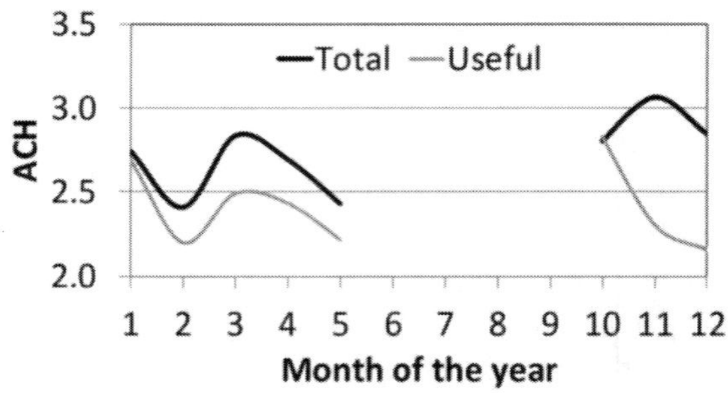

Figure 5. Airflow rate of the classroom.

The solar radiation does not seem to have a noticeable influence on the temperature change at the outlet. The air is an important element to deliver heat in the classroom. Figure 5 shows the air flow rate obtained in the classroom, all over the heating season.

Total, represents the ACH regardless in which the stream direction was obtained. Useful, represents the ACH that entered the classroom upwards, meaning that was pre-heated between the windows. The difference between both data means that were reversed airflow, which was not pre-heated. Anyway, the average season ACH was 2.7. The useful airflow rate, that was heated, has reached an average of about 2.4 ACH. The highest average airflow rate was obtained in November of about 3.1 ACH with a useful airflow rate of about 2.3 ACH. The lowest average airflow rate was obtained in February and May of about 2.4 ACH with a useful airflow rate of about 2.2 ACH. The difference between total and useful airflow rate indicate that a reverse airflow has occurred. This means that the air entered through the chimney, entered the classroom from the ceiling, has crossed it and was exhausted through the ventilated double window. As the outdoor ambient temperature is lower than inside, is the wind effect that influences the ventilation, especially if it is under strong wind conditions.

A more detailed analysis is done, based on an hourly data. For this purpose four hypothetical school weeks were chosen. The first represents the week with highest outdoor air temperature, the second the lowest outdoor air temperature, the third highest global solar radiation and the fourth the lowest global solar radiation. Figure 6 shows the airflow rate of the classroom for these four weeks.

During some periods of time the airflow rate has a negative value. That represents a reversed air motion to what it is desirable, so the air could be warmed up.

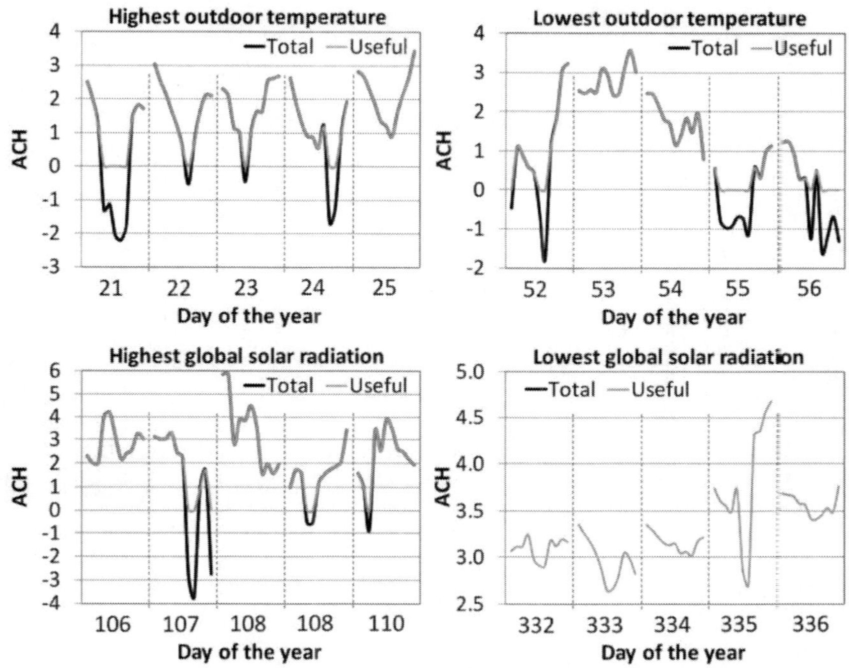

Figure 6. Hourly airflow rate of the classroom during four weeks.

The ventilation is driven by buoyancy and wind effects. Buoyancy is caused by the temperature difference between inside and outside, which tends to make the air rise up when the indoor temperature is higher. Under this circumstance the air is exhausted through the chimney. The wind effect creates a pressure difference between the air inlet (window) and the air outlet (chimney) of the building's envelope. On unfavorable wind conditions, it may not necessarily lead to increased ventilation. The air motion is the result of pressure difference due to buoyancy and wind effect. The first is highly related to the temperature difference and the second to wind conditions. Table 3 shows the wind conditions during the periods from Figure 6 that lead to the highest reverse airflow rate.

During these periods the wind driven pressure difference was higher than the buoyancy pressure difference. The wind acting on the school building creates differences of pressure around it. The wind pressure is positive on the windward side of a building and negative on the leeward side.

Table 3. Wind conditions on reverse airflow rate

Day	Hour	Wind speed (m/s)	Wind direction (degrees)	ACH
21	14	3.8	45	-2.2
24	15	2.4	54	-1.7
52	14	6.1	244	-1.8
56	15	3.1	6	-1.6
107	15	6.8	261	-3.7

Air enters through the windward openings, and leaves through the leeward openings. Wind-induced pressure difference on building surfaces caused by the wind (ΔP_{wind}, in Pa) may be expressed as:

$$\Delta P_{wind} = 0.5(Cp_1 - Cp_2)\rho v_{wind}^2 \tag{2}$$

where, Cp_1 and Cp_2 are pressure coefficients due to wind at the openings 1 and 2 (inlet and outlet of the building), respectively, whose values are in agreement with literature [42], ρ is the density of the air, (kg/m^3) and v_{wind} is the mean wind velocity (m/s). The average useful airflow rate on each week is of 1.5, 1.3, 2.3 and 3.3 ACH for the weeks of the highest outdoor air temperature, the lowest outdoor air temperature, the highest global solar radiation and the lowest global solar radiation, respectively. During the periods shown in Table 3 the wind pressure difference was negative on the window and combined with high wind speed has reversed the direction of the airflow, being then exhausted through the window. As the airflow was reversed the air entered into the classroom at outdoor air temperature. When the air was pre-heated it entered the classroom higher than the outside temperature. Figure 7 shows the hourly air temperatures (outdoor and at the outlet) and the global solar radiation.

Every time the air entered through the windows, the airflow, temperature had been always higher than the outdoor air temperature. The lowest temperature rise is observed during the first week (highest outdoor temperature) is on day 25, at 18:00 h with a rise of 0.9 ºC and the high rise on day 21 at 8:00 h of about 5.4 ºC.

On the day 24 the outlet temperature has risen above the set up the temperature of the classroom (22ºC), up to 25.8 ºC.

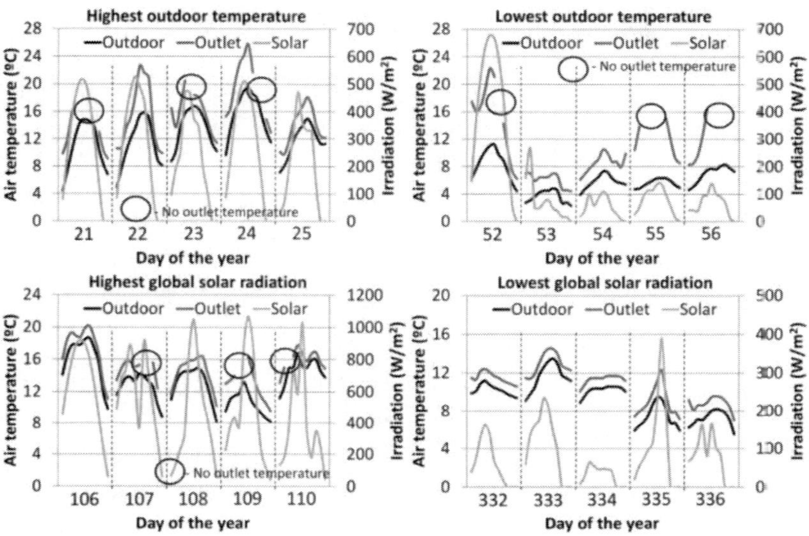

Figure 7. Hourly temperatures and solar irradiation during four weeks.

When the outdoor temperature was lower (second week) the lowest temperature rise is observed on day 52 at 17:00 h, having raised only 1.8ºC and the highest rise is on day 55 at 9:00 of 9ºC. On day 52 at 12:00 h the outlet temperature has risen up to comfort level. Is observed in this second week the overall highest risen of the outlet temperature, when the outdoor air temperature was lower. During the third week (highest global solar radiation) the lowest temperature rise is observed on day 108 at 9:00 of 0.6ºC and the highest rise is on day 109 at 13:00 h of 4.5ºC. The lowest temperature rise during the last week (lowest global solar radiation) is on day 333 at 14:00 of 1ºC. The highest temperature rise is on day 336 at 8:00 h of 3ºC. The temperature rise on these two weeks never reached the classroom temperature of 22ºC. As said before, the solar radiation does not influence much the rise of the delivered air temperature. On the contrary, the airflow has a great influence on the delivered air temperature. On the days 56 and 336, at 12:00 h the outdoor air temperature is 7.5ºC. The global solar radiation is similar, of 97 and 79 W/m^2, respectively, on day 56 and 336. In spite of these similarities the delivered air temperature reached up to 16.3 ºC on the day 56 and only 8.9ºC on the day 336. The airflow rate obtained on those two moments was 0.3 ACH on the day 56 and 3.6 ACH on the day 336. This means that the stream was at a lower speed in the space between the two windows on the day 56. The air temperature increment decreases when air flow rate increases and it increases when airflow rate decreases, as found in another studies [44, 45].

Table 4. Data for the highest useful airflow rate on each week

Day	ACH	Outdoor Temperature (ºC)	Delivered Temperature (ºC)	Temperature Rise (ºC)
25	3.4	11.3	12.2	0.9
53	3.6	2.8	4.6	1.8
108	5.9	11	11.7	0.7
335	4.7	5.9	7.1	1.2

This is because the total heat collected by the air stream from the glazing surfaces depends on the time in which air remains in contact with the hotter surfaces. Comparing figures 6 and 7, it can be seen the highest useful airflow rate that occur in each week. Table 4 shows the data of such days.

From the previous Table it is seen that the lowest rise temperature of the delivered air is on day 25, with the lowest airflow rate, but with the highest outdoor air temperature. However, the highest rise temperature of the delivered air is on day 53, with the lowest outdoor air temperature. Knowing the indoor and outdoor air temperatures and also the volumetric airflow rate the heat loss through the ventilation air (Q_{air}, in W) may be determined as follow:

$$Q_{air} = C\rho V(\theta_{ind} - \theta_{out}) \quad (3)$$

where, C is the specific heat of the air at constant pressure (J/kgK), V is the volumetric air flow rate (m^3/s), θ_{ind} and θ_{out} are the indoor and outdoor air temperatures (ºC), respectively (ºC). Also knowing the outlet air temperature when there is a pre-heating system like the one under analysis, the useful heat (Q_{use}, in W), obtained by the deliverd air, for each flow rate can be determined as:

$$Q_{use} = C\rho V(\theta_{del} - \theta_{out}) \quad (4)$$

where, θ_{del} is the delivered air temperature (ºC). The thermal performance of this building component must compare the heat gain and the heat loss, which result in the heat balance. The ventilated double window has also heat gain due to direct solar transmission to the classroom, through the glazing. The heat loss is the result of the the heat that crosses the inner window to the air channel formed between the two windows.

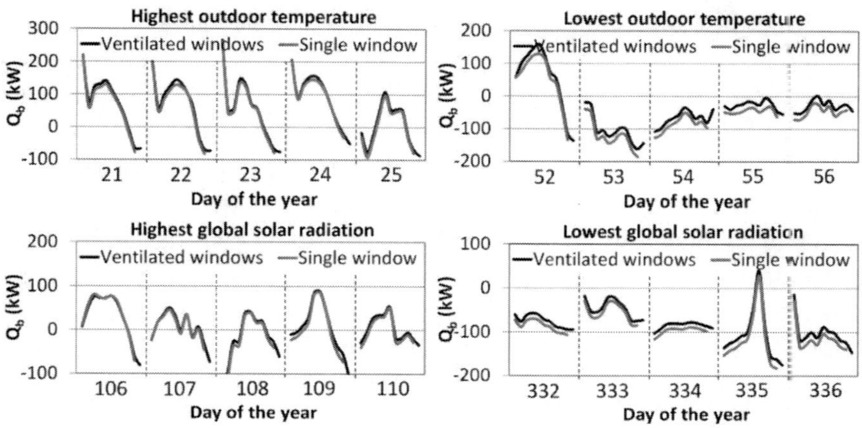

Figure 8. Heat balance of the ventilated double window and the single window including ventilation.

Comparing the ventilated double window and the existing single window among natural ventilation without pre heating, Figure 8 shows the estimated heat balance (Q_b, in W) for the same airflow rate.

The heat balance was estimated through the following equations. For the ventilated double window:

$$Q_b = Q_{use} + Q_{sun} - Q_{loss} - Q_{air} \tag{5}$$

And for the single window:

$$Q_b = Q_{sun} - Q_{loss} - Q_{air} \tag{6}$$

where, Q_{sun} is the solar heat gain (w), Q_{loss} is the heat loss through the window (W). The Q_{sun} in the ventilated double window refers only to the direct solar transmittance to the classroom, once the radiated and convected solar heat flow resulted from the portion of solar radiation absorbed by the windows is included in the Q_{use}. In the single window the Q_{sun} is the sum of a directly transmitted radiant flow fraction and a secondary inward heat flow fraction. The latter corresponds to the portion of solar radiation absorbed by the window and then radiated and convected into the classroom.

As it can be seen from Figure 8 the Q_b of the ventilated double window is mostly higher than of the single window. For instance on the day 106, the combination of high global solar radiation and high outdoor air temperature has inversed the comparison for the Q_b between both systems. First of all, ths single window allows more direct solar radiation into the interior and secondly the ventilated double window did not recovered much heat from inside due to high ambient temperature. Under these conditions there is lower heat loss through the inner window that can be recovered by the passing air. It can be observed that the heat balance at the moments with the highest solar radiation is positive which means that there is in fact a heat gain due to solar radiation. The ventilated double window, at high solar radiation provides an air that is warmer than the indoor air and can contribute to the heating of the indoor environment. The percentage of heat loss that is compensated with the pre-heating of the ventilation air can be calculated as:

$$\frac{Q_{use}}{Q_{air}} 100\% \quad (7)$$

Table 5 shows the maximum, mean and minimum of the heat loss that is compensated with warmer ventilation air.

During the four weeks, the heat loss through the ventilation was compensated from 7% to 243%. The last week presents the lowest values, with a mean heat loss compensation of 11%. This was the week with lowest global solar radiation and also a low ambient temperature. If a lower temperature allows the ventilated double window to recover more heat that is lost from inside, the heat loss through the exhausted air is also higher. The first week is the one that presents higher values, reaching a mean value of 43% of heat loss compensation. This was a week with higher outdoor air temperature, thus a lower heat loss through the ventilated air. In spite of different climatic conditions, these results show that this passive air heating system can recover heat loss through it that, otherwise, could be lost to the outside.

In buildings designed to exploit natural ventilation, human thermal comfort must be considered. Air velocity must be considered as a variable.

Table 5. Percentage of heat loss compensated with warm air (%)

	1st Week	2nd week	3rd week	4rd week
Maximum	243	102	46	24
Mean	43	27	21	11
Minimum	9	10	7	8

The factors are based on ISO 7730:1994 Standard [41]. For buildings without significant air movement, it is recommends temperatures within 20 – 24°C provided that the maximum velocities do not exceed 0.2 m/s. However, if natural ventilation induces higher air speeds, the maximum temperature of 24°C provided that the air velocities do not exceed about 0.4 m/s. The air speed may be found through the following equation:

$$v = \frac{V}{A} \tag{8}$$

where, v is the air velocity (m/s), V is the volumetric air flow rate (m^3/s) and A, is the section área of the pathway air (m^2). The highest atmospheric air speed entering through the outlet of the reference pre-heating system was 1.9 m/s on day 108 at a minimum temperature of 11.7°C. This is the extreme conditions verified during the four weeks under study. Due to the classroom section the air velocity is estimated of about 0.03 m/s. Overall, for this particular case, the air moves with an average velocity between 0.01 and 0.02 m/s throughout the classroom.

Conclusion

To evaluate the energy efficiency of the school building during winter time, pre-heating the incoming air is under consideration. The proposed method is to transform the single large windows of the classrooms into a ventilated double window, which takes advantage of the thermal losses from inside and also on the incident solar radiation. The natural airflow can be introduced due to wind around the building and by thermal gradients between the building and the atmosphere. This study investigated the effect of pre-heating air on a cross ventilation flow regime in a public school building. The first priority of school buildings must be to avoid classroom conditions that make the learning process more difficult due to health problems. Saving energy is in this context a secondary consideration. This means that conserving energy in schools by reducing ventilation is counterproductive, in the sense that they decrease school performance even though they save energy. School buildings, as other buildings, must be designed to run in a sustainable way, augmenting the IAQ and at the same time energy consumption reduction should be sought. This is possible using passive strategies as a ventilated double window. Lack of appropriate natural ventilation is often responsible for pollutant exposure and health symptoms in schools. Portugal has no statistical indicators showing that schools have high rates of respiratory symptoms. According to the literature, poor IAQ in classrooms is associated to CO_2 concentration. Thus, adequate ventilation should be a major focus of school design.

The kind of elementary school from the "Plano dos Centenários" sought an architecture which had a relationship with its surroundings and the location, with a new approach to the solar radiation resource. This was a new approach to solar capture and also to natural ventilation. Numerical simulations were performed to examine the natural ventilation effectiveness of a classroom and the design influences of the ventilated double window. The combined effects of the weather conditions were studied during the heating season. The investigated building proved to be very sensitive to the wind change effects on the natural ventilation. Although, in this work the air pathway is uprising, in other case it could be reversed. It depends on the wind direction relative to the outlet of the chimney duct, which is located above the roof and the inlet on the window. A defined schedule was used for the simulations that were carried out to analyse the performance of a ventilated double window on an elementary school of Portugal. A hypothetical sealed classroom and its geometry were fixed variables to expect an uprising crossing airflow through the classroom. The vents' geometry were found so the mean season airflow rate would be as needed for the classroom population. However, this flow rate and its pathway may change during the day. One possibility a reverse airflow pathway due to the wind, as seen during the simulations. Other factors related to the use of the building may lead to an airflow change. Opening and closing the door between the classroom and the circulation may change the amount of the airflow and its pathway. This door is usually used on early morning and late afternoon. It is also open during the break times, like during the morning and afternoon periods and also at lunch time.

It can be stated that natural ventilation is capable of ensuring appropriate IAQ inside school buildings during the heating season. Ventilation rates under weather conditions depend on the opening cross section, but also strongly on the location of the openings and their geometry. It is also concluded, that there is a necessity for control of the outlets of the window to prevent overcooling of the building during the heating season. The area of the opening has great influence on a building's energy consumption. As the opening area increases, the airflow increases and also the heating load. The ventilated double window with a controlled vent is an effective way of pre heating ventilation air in school buildings. Using existing construction methods and practices, this passive system requires little labour, and operation to function, being a low cost method of saving heating energy for schools in a cold climate. Regarding thermal comfort, the induced air motion of about 0.03 m/s at the living level, it can satisfy the occupants as with indoor temperature of about 22ºC. Moreover, the incoming air, under natural ventilation, enters the building at outdoor temperature. Besides the energy consumption needed to heat up the arriving air, thermal discomfort may be felt by users. Pre-heating the ventilation air before it enters the building, by this passive technique, is one of the possible strategies to enhance thermal comfort.

This double window system is able to provide preheated ventilation air at winter time, by recovering part of the heat losses from indoors and by collecting part of the absorbed solar radiation by the windows. However, solar radiation plays a relatively modest role in the performance of the ventilated double window. In fact, most of the incident radiation is directly transmitted inside through the transparent glazing. Simulating this window system, it was found that the higher the air temperature differences between indoors and outdoors, the higher the heat gained by the delivered air through the system. The ventilation schedule that was used during this study is related to the building use and to comply with the needs of the inhabitants' health. To reduce heat loss through ventilation, an operable vent (windows' outlet) can be opened and closed accordingly to the human needs by reducing or increasing the incoming air flow rate. However, other pollutant agents should be removed, as dust, chemicals from cleaning products, building materials and others. Thus, a continuous airing should go on through all day and night, even on a reduced rate. Finally, it is our opinion that natural ventilation and energy consumption reduction seems to be feasible and viable. Understanding this window system behaviour will provide insight for further research to increase the solar collection capacity. This, of course, awaits a large scale testing in order to see how effective such a ventilation system would be in a real building.

NOMENCLATURE

A	Area, m^2
C	Specific heat, J/kgK
Cp	Pressure coefficients
P	Pressure, Pa
Q	Heat, W

v	Velocity, m/s
V	Volumetric air flow rate, m^3/s
θ	Temperature, ºC
ρ	Density, kg/m^3

Subscript

air	Air
b	Balance
del	Delivered air
ind	Indoor
loss	Loss
out	Outdoor
sun	Sun
use	Useful
wind	Wind

REFERENCES

[1] M. do C. Freitas, N. Canha, M. Martinho, M. Almeida-Silva, S. M. Almeida, P. Pegas, C. Alves, C. Pio, M. Trancoso, R. Sousa, F. Mouro, and T. Contreiras, "Indoor Air Quality in Primary Schools," in Advanced Topics in Environmental Health and Air Pollution Case Studies, InTech, 2011, pp. 361–384.
[2] J. M. Daisey, W. J. Angell, and M. G. Apte, "Indoor air quality, ventilation and health symptoms in schools: an analysis of existing information," *Indoor Air*, vol. 13, no. 1, pp. 53–64, Mar. 2003.
[3] D. Martin, "Are schools making kids sick?," 2012.
[4] P. Wargocki and D. P. Wyon, "Providing better thermal and air quality conditions in school classrooms would be cost-effective," *Building and Environment*, vol. 59, pp. 581–589, Jan. 2013.
[5] M. Turunen, O. Toyinbo, T. Putus, A. Nevalainen, R. Shaughnessy, and U. Haverinen-Shaughnessy, "Indoor environmental quality in school buildings, and the health and wellbeing of students," *International Journal of Hygiene and Environmental Health*, vol. 217, no. 7, pp. 733–739, Sep. 2014.

[6] P. N. Pegas, M. G. Evtyugina, C. A. Alves, T. Nunes, M. Cerqueira, M. Franchi, C. Pio, S. M. Almeida, and M. do C. Freitas, "Outdoor/indoor air quality in primary schools in Lisbon: a preliminary study," *Química Nova*, vol. 33, no. 5, pp. 1145–1149, 2010.

[7] S. Fraga, E. Ramos, A. Martins, M. J. Samúdio, G. Silva, J. Guedes, E. O. Fernandes, and H. Barros, "Indoor air quality and respiratory symptoms in Porto schools," *Revista Portuguesa de Pneumologia*, vol. 14, no. 4, pp. 487–507, 2008.

[8] A. S. for H., Refrigerating and Air Conditioning Engineers ASHRAE, Ventilation for Acceptable Indoor Air Quality. 1999.

[9] D. Teli, M. F. Jentsch, and P. A. B. James, "Naturally ventilated classrooms: An assessment of existing comfort models for predicting the thermal sensation and preference of primary school children," *Energy and Buildings*, vol. 53, pp. 166–182, Oct. 2012.

[10] D. Teli, P. A. B. James, and M. F. Jentsch, "Thermal comfort in naturally ventilated primary school classrooms," *Building Research and Information*, vol. 41, no. 3, pp. 301–316, Jun. 2013.

[11] C. E. de N. CEN, Indoor environmental input parameters for design and assessment of energy performance of buildings- addressing indoor air quality, thermal environment, lighting and acoustics. 2007.

[12] R. M. S. F. Almeida and V. P. de Freitas, "Indoor environmental quality of classrooms in Southern European climate," *Energy and Buildings*, vol. 81, pp. 127–140, Oct. 2014.

[13] Jorge Manuel da Silva Carlos, "Escolas do ensino básico 'tipificadas' (Plano dos Centenários) - avaliação do seu desempenho térmico e propostas de intervenção," Universidade de Coimbra, Coimbra, 2005.

[14] T. G. Theodosiou and K. T. Ordoumpozanis, "Energy, comfort and indoor air quality in nursery and elementary school buildings in the cold climatic zone of Greece," *Energy and Buildings*, vol. 40, no. 12, pp. 2207–2214, Jan. 2008.

[15] O. S. Asfour and M. B. Gadi, "A comparison between CFD and Network models for predicting wind-driven ventilation in buildings," *Building and Environment*, vol. 42, no. 12, pp. 4079–4085, Dec. 2007.

[16] A. Bastide, P. Lauret, F. Garde, and H. Boyer, "Building energy efficiency and thermal comfort in tropical climates: Presentation of a numerical approach for predicting the percentage of well-ventilated living spaces in buildings using natural ventilation," *Energy and Buildings*, vol. 38, no. 9, pp. 1093–1103, Sep. 2006.

[17] A. Aumann, "School buildings on the path to the 2000-Watt Society," *International Journal of Sustainable Building Technology and Urban Development*, vol. 3, no. 1, pp. 60–67, Mar. 2012.
[18] T. Schulze and U. Eicker, "Controlled natural ventilation for energy efficient buildings," *Energy and Buildings*, vol. 56, pp. 221–232, Jan. 2013.
[19] J. S. Carlos and H. Corvacho, "Ventilated Double Window for the Preheating of the Ventilation Air Comparison of Its Performance in a Northern and a Southern European Climate," *Journal of Renewable Energy*, vol. 2013, pp. 1–11, 2013.
[20] J. S. Carlos, "Assessment of pre-heating air through a double window system on different building location and weather condition," *Building Simulation*, vol. 7, no. 3, pp. 247–261, Jun. 2014.
[21] E. Gratia and A. De Herde, "Is day natural ventilation still possible in office buildings with a double-skin façade?," *Building and Environment*, vol. 39, no. 4, pp. 399–409, Apr. 2004.
[22] W. You, M. Qin, and W. Ding, "Improving building facade design using integrated simulation of daylighting, thermal performance and natural ventilation," *Building Simulation*, vol. 6, no. 3, pp. 269–282, Sep. 2013.
[23] K. Gładyszewska-Fiedoruk and A. Gajewski, "Effect of wind on stack ventilation performance," *Energy and Buildings*, vol. 51, pp. 242–247, Aug. 2012.
[24] A.-S. Yang, C.-Y. Wen, Y.-H. Juan, Y.-M. Su, and J.-H. Wu, "Using the central ventilation shaft design within public buildings for natural aeration enhancement," *Applied Thermal Engineering*, vol. 70, no. 1, pp. 219–230, Sep. 2014.
[25] J. Khedari, B. Boonsri, and J. Hirunlabh, "Ventilation impact of a solar chimney on indoor temperature fluctuation and air change in a school building," *Energy and Buildings*, vol. 32, no. 1, pp. 89–93, Jun. 2000.
[26] S. R. Livermore and A. W. Woods, "Natural ventilation of multiple storey buildings: The use of stacks for secondary ventilation," *Building and Environment*, vol. 41, no. 10, pp. 1339–1351, Oct. 2006.
[27] P. Lourenço, M. D. Pinheiro, and T. Heitor, "From indicators to strategies: Key Performance Strategies for sustainable energy use in Portuguese school buildings," *Energy and Buildings*, vol. 85, pp. 212–224, Dec. 2014.
[28] J. S. Carlos and H. Corvacho, "Retrofit measures in old elementary school buildings towards energy efficiency," *Journal of Civil Engineering and Management*, vol. 16, no. 4, pp. 567–576, Dec. 2010.

[29] E. G. Dascalaki and V. G. Sermpetzoglou, "Energy performance and indoor environmental quality in Hellenic schools," *Energy and Buildings*, vol. 43, no. 2–3, pp. 718–727, Feb. 2011.

[30] L. de Santoli, F. Fraticelli, F. Fornari, and C. Calice, "Energy performance assessment and a retrofit strategies in public school buildings in Rome," *Energy and Buildings*, vol. 68, pp. 196–202, Jan. 2014.

[31] D. A. Krawczyk, "Theoretical and real effect of the school's thermal modernization—A case study," *Energy and Buildings*, vol. 81, pp. 30–37, Oct. 2014.

[32] A. W. Woods, S. Fitzgerald, and S. Livermore, "A comparison of winter pre-heating requirements for natural displacement and natural mixing ventilation," *Energy and Buildings*, vol. 41, no. 12, pp. 1306–1312, Dec. 2009.

[33] H. Manz, H. Huber, A. Schälin, A. Weber, M. Ferrazzini, and M. Studer, "Performance of single room ventilation units with recuperative or regenerative heat recovery," *Energy and Buildings*, vol. 31, no. 1, pp. 37–47, Jan. 2000.

[34] F. Peci López and M. Ruiz de Adana Santiago, "Sensitivity study of an opaque ventilated façade in the winter season in different climate zones in Spain," *Renewable Energy*, vol. 75, pp. 524–533, Mar. 2015.

[35] R. C. Richman, C. Cianfrone, and K. D. Pressnail, "More Sustainable Masonry Facades: Preheating Ventilation Air Using a Dynamic Buffer Zone," *Journal of Building Physics*, vol. 34, no. 1, pp. 27–41, Jul. 2010.

[36] J. S. Carlos, H. Corvacho, P. D. Silva, and J. P. Castro-Gomes, "The contribution of a double ventilated window to the energy efficiency of a building," in *Advances in Energy Research*, vol. 7, New York: Nova Science Publishers, Inc., 2011, pp. 273–306.

[37] J. S. Carlos, H. Corvacho, P. D. Silva, and J. P. Castro-Gomes, "Modelling and simulation of a ventilated double window," *Applied Thermal Engineering*, vol. 31, no. 1, pp. 93–102, Jan. 2011.

[38] J. O. Lewis, J. R. Goulding, and T. C. Steemers, Energy in Architecture: The European Passive Solar Handbook. London, UK: B. T. Batsford, 1992.

[39] European Committee for Standardization, Thermal performance of buildings - Calculation of energy use for heating - Residential buildings. 1998.

[40] E. Hatzidimoula and K. Poulios, "Natural ventilation of indoor spaces while minimizing heat losses: Theoretical and experimental

investigation," *Energy and Buildings*, vol. 40, no. 4, pp. 621–626, Jan. 2008.

[41] (ISO) International Organization for Standardization, Moderate thermal environments - Determination of the PMV and PPD indices and specification of the conditions for thermal comfort. 1994.

[42] European Commission and ALTENER Programme, Natural ventilation in buildings: a design handbook. London: James and James (Science Publishers) Ltd, 1998.

[43] J. S. Carlos, H. Corvacho, P. D. Silva, and J. P. Castro-Gomes, "Heat recovery versus solar collection in a ventilated double window," *Applied Thermal Engineering*, vol. 37, pp. 258–266, May 2012.

[44] M. McEvoy, R. Southall, and P. Baker, "Test cell evaluation of supply air windows to characterise their optimum performance and its verification by the use of modelling techniques," *Energy and Buildings*, vol. 35, no. 10, pp. 1009–1020, Nov. 2003.

[45] J. S. Carlos, H. Corvacho, P. D. Silva, and J. P. Castro-Gomes, "Real climate experimental study of two double window systems with preheating of ventilation air," *Energy and Buildings*, vol. 42, no. 6, pp. 928–934, Jun. 2010.

In: Natural Ventilation
Editor: Oscar T. Haynes
ISBN: 978-1-63482-414-9

Chapter 5

EVALUATION OF THERMAL COMFORT CONDITIONS IN SPACES EQUIPPED WITH RADIANT FLOOR SYSTEM

Eusébio Z. E. Conceição
Faculty of Sciences and Technology, University of Algarve, Faro, Portugal

ABSTRACT

The aim of this chapter is to evaluate numerically the comfort conditions in spaces equipped with radiant floor system and forced ventilation system. The results of this original chapter are obtained in a virtual chamber equipped with a desk, two seats and two virtual mannequins. In this work the numerical softwares, developed by the author, are presented.

In this numerical work the thermal comfort, local thermal discomfort and air quality levels are evaluated in two occupants in a space equipped with radiant floor system and forced ventilation system. This study, developed from three coupled software, considers the buildings thermal response, computational fluids dynamics and human thermo-physiology models.

In this simulation the temperature field on the building surfaces and collector water are calculated from the buildings thermal response with complex topology, the field of air temperature and air velocity around the occupants is calculated from the computational fluids dynamics and the temperature field in the occupants and clothing is calculated from the human thermo-physiology models.

INTRODUCTION

In the presented study the thermal comfort conditions, local thermal discomfort and air quality in a virtual chamber equipped with forced ventilation and radiant floor system, in winter conditions, promoted by a solar collector, are evaluated.

The idea of this study is to develop and apply three numerical software, based on the coupling of two integral and one differential models in order to study combinations of different conditions of forced ventilation and radiant floor system.

In the forced ventilation, outside air environment is used, while in the radiant floor water system a solar collector is used. The combination of the two systems, with the presence of occupants, is used to promote acceptable thermal comfort, local thermal discomfort and air quality conditions, with low energy consumption level.

In the numerical study the integral building thermal response, the integral human thermo-physiology and the differential computational fluids dynamics are used.

The study presented in this paper represents a continuation of Conceição et al. (2006), Conceição et al. (2008a), Conceição et al. (2008b), Conceição et al. (2010a), Conceição et al. (2010b) and Conceição et al. (2012). Other studies of ventilation in interior spaces can be analyzed, for example, in Kobayashi et al. (2007), Nonaka et al. (2007), Sawachi et al. (2009), Imano et al. (2009), Tsukamoto et al. (2009), Kobayashi et al. (2009) and Kouchi et al. (2009). The application of the coupling integral and differential models can be found, in some applications, for example, in Gau et al. (2006), Zhu et al. (2007) and Omni et al. (2007).

In the evaluation of thermal comfort levels the PMV indexes (Predicted Mean Vote) and PPD (Predicted Percentage of Dissatisfied people) are used (see Fanger, 1970, ISO 7730, 2005 and ANSI/ASHRAE 55, 2004). These indexes are influenced by four environmental variables (air temperature, air velocity and air relative humidity and mean radiant temperature), that each section of the human body is subjected, and two personal parameters (clothing level and physical activity).

The local thermal discomfort level can be evaluated using the Draught Risk. The Draught Risk, which depends on the air temperature, air velocity and air turbulence intensity, developed by Fanger et al. (1988), was applied to several works such as, for example, Fanger and Christensen (1986) and Melikov et al. (1990).

Finally, in order to access the air quality level is used the carbon dioxide concentration, released by the occupants, evaluated by the differential computational fluids dynamics.

The recommendations of air quality level can be seen in international regulations such as ANSI/ASHRAE Standard 62.1 (2004) and national portuguese regulations such as D-L N. 79/2006 of 4 April. Some applications of carbon dioxide concentration released by the occupant, used as an indicator of indoor air quality, can be found, for example, in Papakonstantinou et al. (2002), Nagy and Senitkova (2007), Bulinska (2007) and Levine (1993).

Numerical Models

This work used the coupling of numerical models that simulate the integral buildings thermal response (Conceição, 2003, Conceição and Lúcio, 2008 and Conceição and Lúcio, 2006), the differential computational fluids dynamics (see Conceição et al. 2008b and Conceição et al. 2010b) and the integral human thermal-physiology (Conceição et al., 2006, Conceição et al., 2006, Conceição et al., 2010a and Conceição et al., 2010b). The integral buildings thermal response is used for assessing the temperatures of the transparent and opaque building surfaces, the indoor environmental variables and the buildings surfaces solar radiations. The differential computational fluids dynamics (Patankar, 1980) is not the only method used to assess the airflow around the occupants (temperature and air velocity, turbulence intensity and Draught Risk) (Popiolek and Melikov, 2004), but also to assess the level of indoor air quality (using the carbon dioxide concentration).

The integral human thermal-physiology (Stolwijk's, 1970) is used to evaluate the thermal comfort (Fanger, 1970), the body temperature, the clothes temperature and water mass in the skin surface and clothing.

Numerical Methods

In the numerical simulation is used a virtual compartment equipped with:

- Radiant floor system;
- Forced ventilation system, with the inlet and outlet located on adjacent walls, located above the level of the head of the occupants;

- One desk and two seats;
- Two seated virtual occupants.

In table 1 are presented the parameters used in the simulation.

In the computational fluids dynamics is used a grid generation of 48x54x48, with 5.45 cm spacing's in the X axis direction, 5.2 cm spacing's in the Y axis direction and 5.45 cm spacing's in the Z axis direction.

In the human thermo-physiology and computational fluids dynamics numerical models, the occupant has 1.70 m height, 70 Kg weight, 1.2 Met activity and 1 Clo clothing level.

RESULTS AND DISCUSSION

Temperature Field in the Virtual Chamber

In table 2 the surfaces temperature field in the virtual chamber, calculated by the building thermal numerical model are presented.

Table 1. Definition of the parameters of each of the four situations studied

	Outdoor Air Temperature (°C)	Collector area (m^2)	Airflow rate (m^3/s)
Situation I	16	2	0,2
Situation II	16	2	0,02
Situation III	10	4	0,02
Situation IV	16	4	0,02

Airflow around the Occupants

It can be concluded that the situations I and III have values of temperature lower than the situations II and IV. In situation I, as a result of the higher air velocity values, air temperature is lower than situations II and IV. In situation III the air temperature values are lower due to the fact that in this situation is imposed an outside temperature value lower than the situations I, II and IV.

Table 2. Temperature field in the virtual camera and solar radiation

	Situation I	Situation II	Situation III	Situation IV
Right surface temperature (°C)	17,423	20,365	17,761	23,726
Left surface temperature (°C)	17,408	20,318	17,678	23,643
Front surface temperature (°C)	17,436	20,403	17,829	23,794
Back surface temperature (°C)	17,405	20,307	17,658	23,624
Ceiling surface temperature (°C)	16,833	18,555	14,543	20,523
Floor wood surface temperature (°C)	27,916	30,233	35,307	41,193
Floor cement surface temperature (°C)	34,758	36,447	46,356	52,192
Floor isolating surface temperature (°C)	21,708	22,155	25,537	28,884
Solar radiation (W/m^2)	850	850	850	850

The inlet air introduced through the forced ventilation system comes from the outside environment, so the inlet air temperature has the same value as the outdoor temperature, which will influence the temperature value inside the compartment.

It is also possible that the values of air temperature in the situations IV are higher than the air temperature values in the situations I, II and III. In the situation IV is imposed a higher value than the collector area, which results in a higher floor temperature influencing the airflow temperature inside the compartment.

It can also be checked the stratification of the air temperature, at least 1 m above ground level, caused by heated radiant surface.

Thermal Comfort Level

In Tables 3 and 4 are shown the results of the PMV and PPD, respectively, of the two occupants, for each of the four conditions studied. Through the PMV and PPD values presented in tables 3 and 4, it can be seen that the situation I is included in category A, the situations II and III are included in the category C and the situation IV the two occupants are not in comfortable conditions, according to ISO 7730 (2005).

Local Thermal Discomfort

On this section is analyzed and discussed the Draught Risk that the two occupants are subjected. In the situation I the Draught Risk values are higher in sections of the feet and legs of the two occupants and, according to ISO 7730 (2005), classified on category C. In situations II, III and IV, the Draught Risk values are low, fitting into the category A of ISO 7730 (2005).

Air Quality

It is possible to verify that in the situations I, carbon dioxide contaminated area is lower than the carbon dioxide concentration observed in situations II, III and IV.

Table 3. PMV values for each of the two occupants in each of the four studied situations

PMV	Situation I	Situation II	Situation III	Situation IV
1	-0,199	0,454	0,173	0,973
2	0,013	0,589	0,731	1,023

Table 4. PPD values of each of the two occupants in each of the four studied situations

PPD	Situation I	Situation II	Situation III	Situation IV
1	5,817	9,306	5,617	24,993
2	5,004	12,264	16,235	27,113

From Table 5 it is possible to check the carbon dioxide concentration in the breathing area of each of the two occupants in each of the four studied situations. It can be checked that situation I gives an acceptable quality of air, which is less than 1800 mg/m^3 (see ASHRAE and Decree-Law). In situations

II, III and IV there is no acceptable quality of air, with the concentration in the breathing area of the occupants higher than 4000 mg/m^3.

CONCLUSION

In this numerical work are evaluated the thermal comfort, local thermal discomfort and air quality levels, that two occupants are subjected in a space equipped with a radiant floor system and a forced ventilation system. This study, developed using three software programs, considers the buildings thermal response, human thermo-physiology and computational fluids dynamics.

Four simulations are done for winter conditions. It is analyzed the effect of different values of airflow rate, solar collector area and outside ambient temperature. In the simulation are considered a virtual chamber, with two seated occupant, equipped with a radiant floor system and a system forced ventilation. In accordance with the results obtained in situation I and in situation II, it can be concluded that the increasing of the airflow rate, increase the air velocity inside the compartment, resulting in a decreasing of the temperature in the space.

Both situations promote acceptable thermal comfort conditions. In the local thermal discomfort the situation I is classified in category C, while in situation II is classified in category A. However, unlike the situation II, the situation I has an acceptable the air.

In accordance with the results obtained in situation II and in situation IV, it can be verified that the increasing of the solar collector area, the indoor air temperature increase and the PMV and PPD values are higher.

Table 5. Concentration of carbon dioxide in the breathing area of each of the two occupants in each of the four situations

Carbon dioxide in the breathing area (mg/m^3)	Situation I	Situation II	Situation III	Situation IV
occupant 1	1257,8	4296,85	4176,51	4377,64
occupant 2	1271,06	4111,727	4026,38	4155,65

The air velocity and Draught Risk are reduced and an unacceptable air quality, with higher carbon dioxide concentrations in the breathing area of the occupants, is verified.

In accordance with the results obtained in situation III and in situation IV, it can be concluded that the increasing of the exterior environment temperature an increasing the indoor air temperature is verified. The situation III presents acceptable values of PMV, which is not the case in the situation IV. Both situations have acceptable Draught Risk, however the air quality is not acceptable.

From the results obtained in this study, can be concluded that higher airflow rate obtained acceptable thermal comfort conditions, local thermal discomfort and air quality, for the two occupants of the space.

The developed numerical tool, applied in this work in radiant floor system and a forced ventilation system, can be used in the future to develop better and more efficient integrated radiant systems in buildings or in winter conditions, with solar collectors or biomass systems, either in summer conditions, with systems based on geothermal or solar air conditioned.

REFERENCES

ANSI/ASHRAE Standard 62.1. 2004. ASHRAE Standard – Ventilation for acceptable indoor air quality, American Society of Heating, Refrigerating and Air-Conditioning Engineers, Inc.

Bulinska, A. 2007. Determination of Airflow Pattern in a Residential Building Using Metabolic Carbon Dioxide Concentration Measurements. *RoomVent'2007, Tenth International Conference on Air Distribution in Rooms*, Helsinki, Finland, 13-15 June 2007.

Conceição, E. Z. E. 2003. Numerical Simulation of Buildings Thermal Behaviour and Human Thermal Comfort Multi-node Models. In: *Proceedings of the 8th International IBPSA Conference - Building Simulation.* Eindhoven, Netherland, 11 to 14 of August 2003, 1: pp. 227-234.

Conceição, E. Z. E. and Lúcio, Mª M. J. R. 2006. Numerical study of thermal of school buildings in summer conditions, In: *Proceedings of the Healthy Buildings*, Lisbon, 3: pp. 195-200.

Conceição, E. Z. E. and Lúcio, Mª M. J. R. 2008. Thermal Study of School Buildings in Winter Conditions. *Building and Environment*, 43(5), pp. 782-792.

Conceição, E. Z. E., Lúcio, Mª M. J. R., Capela, T. L., and Brito, A. I. P. V. 2006. Evaluation of thermal comfort in slightly warm ventilated spaces in non-uniform environments. *International Journal of Heating, Ventilating, Air-Conditioning and Refrigerating Research, ASHRAE*, Vol. 12, N. 3, pp. 451-458.

Conceição, E. Z. E., Rosa, S. P., Custódio, A. L. V., Andrade, R. L., Meira, Mª J. P. A., and Lúcio, Mª M. J. R. 2010a. Study of Airflow Around Occupants Seated in Desks Equipped With Upper and Lower Air Terminal Devices for Slightly Warm Environments, *International Journal on Heating Air Conditioning and Refrigerating Research, ASHRAE*, Vol. 16, N. 4, July 2010, pp. 401-412.

Conceição, E. Z. E., Lúcio, Mª M. J. R., Rosa, S. P., Custódio, A. L. V., Andrade, R. L., and Meira, Mª J. P. A. 2010b, Evaluation of comfort level in desks equipped with two personalized ventilation systems in slightly warm environments, *Building and Environment*, Volume 45, Issue 3, March 2010b, pp. 601-609.

Conceição, E. Z. E., Lúcio, Mª M. J. R., Vicente, V. D. S. R., and Rosão, V. C. T. 2008a. Evaluation of Local Thermal Discomfort in a Classroom Equipped with Crossed Ventilation, *The International Journal of Ventilation,* UK, Vol. 7, Nº 3, December 2008, pp. 267-277.

Conceição, E. Z. E., Vicente, V. D. S. R. and Lúcio, Mª M. J. R. 2008b. Airflow inside school buildings office compartments with moderate environment. *International Journal on Heating Air Conditioning and Refrigerating Research, ASHRAE*, Vol. 14, N. 2, March 2008, pp. 195-207.

Decreto-Lei no. 79. 2006. Regulamento dos sistemas energéticos de climatização em edifícios (RSECE), *Diário da República*, I Série- A, N. 67, 4 de Abril.

Fanger, P. O. 1970. *Thermal comfort*. Copenhagen: Danish Technical Press.

Fanger, P. O. and Christensen, N. K. 1986. Prediction of Draught in Ventilated Spaces, *Ergonomics*, V. 29, pp. 215-235.

Fanger, P., Melikov, A. and Hazawa, H. and Ring, J. 1988. Air turbulence and sensation of draught, *Energy and Buildings*, Vol. 12, pp. 21-39.

Gau, N., Niu, J. and Zang, H. 2006. Coupling CFD and Human Body Thermoregulation Model for the Assessment of Personalized Ventilation. *International Journal of Heating, Ventilating, Air-Conditioning and Refrigerating Research, ASHRAE*, Vol. 12, N. 3, pp. 497-518.

Imano, M., Hoshino, H., Akamine, Y., Kamata, M., Nishimura, A., and Sakamoto, Y. 2009. Validation of CFD Analysis of Cross-Ventilation in

Detached House Uses. *RoomVent'2009, Eleventh International Conference on Air Distribution in Rooms*, Bussan, Korea, 24-27 June 2009.

ISO 7730, 2005. Ergonomics of the thermal environments – analytical determination and interpretation of thermal comfort using calculation of the PMV and PPD indices and local thermal comfort criteria. International Standard. Switzerland.

Kobayashi, T., Sagara, K., Yamanaka, T., Kotani, H., and Sandberg, M. 2007. Prediction of Cross-Ventilation Rate through Large Openings - Problems of Conventional Method. *Room Vent '2007, Tenth International Conference on Air Distribution in Rooms*, Helsinki, Finland, 13-15 June 2007.

Kobayashi, T., Sandberg, M., Kotani, H., Claesson, L., Yamanaka, T., and Sagara, K. 2009. Airflow Patterns and Pressure Distributions for a Cross-Ventilated Detached House Analyzed by Wind Tunnel Tests and CFD Analysis. *RoomVent'2009, Eleventh International Conference on Air Distribution in Rooms,* Bussan, Korea, 24-27 June 2009.

Kouchi, Y., Kurabuchi, T., Ohba, M., Endo, T., and Nonaka, T. 2009. Study on Prediction of Cross-Ventilation Indoor Airflow Distribution with Domain Decomposition Technique. RoomVent'2009, *Eleventh International Conference on Air Distribution in Rooms*, Bussan, Korea, 24-27 June 2009.

Levine, K. B., Sterling, E. M. and Collet, C. W. 1993. Estimation of Outdoor Air Ventilation Rates Using CO_2 Concentration. *ASHRAE Transactions*, Part 1, pp. 1554-1559.

Melikov, A. K., Langkilde, G. and Derbiszewsbi, B. 1990. Airflow Characteristics in the Occupied Zone of Rooms with Displacement Ventilation, *ASHRAE Transactions*, V. 96, Part. 1, pp. 555-563.

Miyanaga, T. and Nakamo, Y. 1998. Analysis of thermal sensation in a radiant cooled room by modified PMV. *Room Vent '1998, Sixth International Conference on Air Distribution in Rooms,* Stockholm, Sweden, V. 2, pp. 125-131.

Miyanaga, T. and Nakamo, Y. 1998. Analysis of Thermal Sensation in a Radiant Cooled Room by Modified PMV. *Room Vent '1998 - 6th International Conference on Air Distribution in Rooms*. KTH, Stockholm, Sweden.

Nagy, R. and Senitkova, K. 2007. IAQ and Distribution Systems in School Buildings. *Room Vent '2007, Tenth International Conference on Air Distribution in Rooms*, Helsinki, Finland, 13-15 June 2007.

Nonaka, T., Kurabuchi, T., Ohba, M., Endo, T., Goto, T., Akamine, Y., and Numa, H. 2007. Investigation Concerning Effect of Roof Window in Cross Ventilation. *Room Vent '2007, Tenth International Conference on Air Distribution in Rooms*, Helsinki, Finland, 13-15 June 2007.

Omni, O. and Tenabe, S. 2007. Coupled Simulation of Convection-Radiation-Thermoregulation for Predicting Human Thermal Sensation. *Room Vent '2007, Tenth International Conference on Air Distribution in Rooms*, Helsinki, Finland, 13-15 June 2007.

Papakonstantinou, K. A., Kiranoudis, C. T. and Markatos, N. C. 2002. Numerical Simulation of CO2 dispersion in an auditorium. *Energy and Buildings*. Vol. 34, pp. 245-250.

Patankar, S. V. 1980. *Numerical heat transfer and fluid flow*. Hemisphere publishing corporation.

Popiolek, Z. and Melikov, A. 2004. Improved interpretation and validation of CFD predictions, *Proceedings of the 9th International Conference Air Distribution in Rooms, Room Vent '2004*, Coimbra, Portugal.

Sawachi, T., Nishizawa, S., Habara, H., and Miura, H. 2009. Estimation of Cooling Energy Reduction by Utilizing Cross-Ventilation in Detached Houses, within the Japanese Newly Introduced Energy Regulation Evaluating Energy Consumption for Different Uses. *RoomVent'2009, Eleventh International Conference on Air Distribution in Rooms*, Bussan, Korea, 24-27 June 2009.

Stolwijk, J. A. J. 1970. Mathematical Model of Thermoregulation. In: Hardy, J. D., Gagge, A. P. and Stolwijk, J. A. J. *Physiological and Behavioral Thermoregulation*. Thomas. Springfield. pp. 703-721. 1970.

Tsukamoto, K., Ohba, M., Kurabuchi, T., Nonaka, T., and Goto, T. Y. 2009. Study on Reduction of Cooling Loads in Detached House by Cross-Ventilation Using Coupled Simulation of Semi-Empirical Ventilation Model and Network Models. *Room Vent '2009, Eleventh International Conference on Air Distribution in Rooms*, Bussan, Korea, 24-27 June 2009.

Zhu, S., Kato, S., Ooka, R., Sakoi, T., and Tsuzuki, K. 2007. Coupled Simulation Method of Convection, Radiation, Moisture Transport and Sakoi Human Thermal Physiological Model to Simulate Heat Exchange from a Person Seated in a uniform environment. *Room Vent '2007, Tenth International Conference on Air Distribution in Rooms*, Helsinki, Finland.

In: Natural Ventilation
Editor: Oscar T. Haynes

ISBN: 978-1-63482-414-9

Chapter 6

APPLICATION OF INTEGRAL AND DIFFERENTIAL VENTILATIONS STRATEGIES IN BUILDINGS SPACES

Eusébio Z. E. Conceição and Mª Manuela J. R. Lúcio
Faculty of Sciences and Technology, University of Algarve, Faro, Portugal

ABSTRACT

The aim of this chapter is to analyse the application of integral and differential ventilations strategies in buildings spaces in the South of Portugal. The results of this review chapter are obtained, in Mediterranean environment, in occupied spaces. In this work the integral and differential numerical strategies and the experimental analysis are considered.

The airflow inside school buildings office compartments with moderate environment, the evaluation of local thermal discomfort in a classroom equipped with crossed ventilation, the evaluation of comfort level in desks equipped with two personalized ventilation systems, the evaluation of thermal comfort conditions in a localized radiant system placed in front and behind two students seated nearby warmed curtains, the evaluation of thermal comfort conditions in a classroom equipped with cooling radiant systems and subjected to uniform convective environment, the application of a developed adaptive model in the evaluation of thermal comfort in ventilated kindergarten occupied spaces, the evaluation of indoor air quality in a classroom equipped with crossed ventilation and the comfort and airflow evaluation in spaces equipped with mixing ventilation and cold radiant floor are some examples of analysed and developed work in this chapter.

1. Building Thermal Response

In this study a numerical software that simulates the building thermal behaviour with complex topology, in transient conditions, is used. The software evaluates the building thermal response, the occupants' thermal comfort and the internal air quality evaluation. The software was previously validated in Summer and Winter real conditions.

The software is based in energy and mass balance integral equations. The equations system is developed through the building geometry using computer design software and the resolution is done by the Runge-Kutta-Felberg with error control. The equations consider the heat convective (natural, forced and mixed), heat conductive, heat radiative (long and short-wave calculus) and mass diffusion phenomena. In the radiative phenomenon, the external and internal shading devices are considered.

The integral buildings thermal response is used for assessing the temperatures of the transparent and opaque building surfaces, the indoor environmental variables (air temperature, velocity and relative humidity and Mean Radiant Temperature) and the buildings solar radiation surfaces.

2. Computational Fluids Dynamics

The computational fluid dynamics model, used in this work, evaluates the airflow around the occupants and inside the space. The software was previously validated in isothermal and non-isothermal conditions.

The computational fluid dynamics model considers the three-dimensional turbulent airflow, in steady-state regime, for non-isothermal conditions. The numerical model considers the Cartesian coordinates and the occupants, desk and seats are simulated automatically.

The numerical model is used to evaluate, among others, the airflow around the occupants (temperature and air velocity, turbulence intensity and draught risk) and the indoor air quality (using the carbon dioxide concentration).

3. Human Thermal-Physiology

A human thermal-physiology numerical model, that allows simulating the human body thermal, clothing thermal and thermo-regulatory systems, in non-uniform environments, is used. The software that works in transient conditions, was previously validated in uniform and non-uniform conditions.

This numerical model is divided in six parts: human body thermal system, clothing thermal system, integral equations resolution system, thermo-regulatory system, heat exchanges between the body and the environment and thermal comfort evaluation.

The thermal system is based on the energy and mass balance integral equations. In the equations system resolution is used the Runge-Kutta-Fehlberg with error control and in the thermal comfort evaluation, in non-uniform environments, is used the Predicted Mean Vote (PMV) and the Percentage of People Dissatisfied (PPD).

The numerical model is used to evaluate the human body temperature, the clothing temperature, the water vapour in the skin and clothing and the thermal comfort that the occupants are subjected.

4. Revision Study

4.1. Cross Ventilation with Inlet and Outlet Placed in Non-Parallel Walls

In Conceição et al. (2008a) the airflow inside school buildings office compartments with moderate environment is evaluated. In the differential study, a computational fluids dynamics software, developed in steady-state conditions, is made.

In this experimental and numerical study is evaluated the thermal comfort, the local thermal discomfort and air quality levels.

The study is made inside an experimental and virtual chamber, with 2.7 × 2.4 × 2.4 m^3, that simulates a school building office. The inlet and outlet airflow is located above the occupants' head level in non-parallel walls.

In this study the numerical model is used to evaluate the mean airflow topology, while the experimental data is used to validate the numerical model and to evaluate the airflow turbulent topology.

In the experimental tests the air velocity fluctuations and the carbon dioxide concentration are measured.

The air velocity fluctuations is used to evaluate the air velocity root mean square, the air turbulence intensity, the draught risk, the air velocity fluctuation frequencies and the air velocity fluctuation equivalent frequencies. The carbon dioxide concentration, using the decay tracer gases method, is used to evaluate the mean and local air renovation rates.

In accordance with the obtained results the airflow is divided into two areas, namely the non-occupied upper area and the occupied lower area. In the first one the air velocity is highest and the contaminant released by the occupants is extracted.

In the second one the air velocity is lowest, the thermal comfort is higher, the local discomfort levels is lower and the contaminant released by the occupants is lower, because is transported to the non-occupied upper area.

4.2. Cross Ventilation with Inlet and Outlet Placed in Parallel Walls

In Conceição et al. (2008b) the evaluation of local thermal discomfort in a classroom equipped with crossed ventilation is made. The experimental study is made in a classroom in a school building, equipped with crossed ventilation, in a moderated Summer typical day, in the South of Portugal with Mediterranean climate. The inlet and outlet airflow is located above the occupants' head level in parallel walls.

In this work three different situations are presented. In each situation is evaluated not only the local thermal discomfort level (using the draught risk and uncomfortable air velocity fluctuations), but also the thermal comfort (using the PMV index), the air quality (using the air renovation rate calculated through the decay tracer gas concentration) and the acoustical comfort levels (using the reverberation time, voice clarity, definition and early reflection ratio). To evaluate the local thermal discomfort levels, the air temperature and velocity fluctuations in different classroom interior points are measured. In the thermal comfort level the air mean temperature, velocity and relative humidity and the Mean Radiant Temperature are measured. In the air quality level the carbon dioxide concentration decay is measured.

Finally, in the acoustical parameters, the noise level decay is measured.

4.3. Personalised Ventilation with Upper and Lower Air Terminal

In Conceição et al. (2010) the evaluation of comfort level in desks equipped with two personalized ventilation systems, in slightly warm environments, is made. The experimental study is made in a wood chamber, with 2.7 × 2.4 × 2.4 m^3. In the numerical simulation a multi-nodal human thermal comfort numerical model and a computational fluid dynamic numerical model are used.

In this study is evaluated the thermal comfort level (using a PMV value calculated by the multi-nodal human thermal comfort numerical model), the local thermal discomfort level (using the draught risk and the air velocity fluctuation equivalent frequencies, calculated by the empirical models) and the air quality level (using the carbon dioxide concentration, calculated by the computational fluid dynamic numerical model).

In the experimental tests a manikin and a ventilated classroom desk are used. In the measuring equipment two indoor climate analysers are used. The air velocity and temperature are measured around 15 human body sections, while the Mean Radiant Temperature and the air relative humidity are measured in the non-occupied area.

The classroom desk is occupied by a student, located in the right side seat, and the personalized ventilation system is equipped with one upper air terminal device located above the desk writing area, in front to the trunk area, and one lower air terminal device located below the desk writing area, in front to the legs area.

4.4. Localized Radiant System Placed in Front and Behind Two Students

In Conceição and Lúcio (2010) the evaluation of thermal comfort conditions in a localized radiant system placed in front and behind two students seated nearby warmed curtains is made. In this numerical study the human thermo-physiology model is used.

The study is made inside a virtual chamber, with 2.7 × 2.4 × 2.4 m^3, equipped with two virtual students, a classroom desk (with a localized radiant system placed in front and behind the occupants) and a window equipped with curtains subjected to solar radiation.

In the first part of the work the Mean Radiant Temperature methods are evaluated. In the second parte the influence of air temperature, the localized radiant system temperature, the lateral curtains temperature and the air relative humidity, that the two occupants are subjected, in the thermal comfort level is made.

4.5. Classroom Cooling Radiant Systems with Uniform Convective Environment

In Conceição and Lúcio (2011) the evaluation of thermal comfort conditions in a classroom equipped with cooling radiant systems and subjected to uniform convective environment is made. The study is made in a virtual classroom, with 24 students and 1 teacher, equipped with 13 desks and 25 seats. In the numerical simulation, made with lightly warm Summer conditions, the human thermal-physiology numerical model is used. In the thermal comfort evaluation the PMV (Predicted Mean Vote) is used.

Seven different radiant systems are combined to three convective environments. The internal air temperature and internal surfaces temperature are 28 ºC, the radiant cooling surfaces temperatures are 19 ºC and the warmed internal curtains surfaces temperatures, subjected to direct solar radiation, are 40 ºC.

4.6. Natural and Forced Ventilation in a Kindergarten

In Conceição et al. (2012a) the application of a developed adaptive model in the evaluation of thermal comfort in ventilated kindergarten occupied spaces is made. The experimental study is made in a kindergarten and is used the building thermal behaviour software. The study is made in real conditions, in Mediterranean environment, for cold and warm thermal conditions.

To evaluate the thermal comfort level, using the PMV index, is developed an adaptive model. The adaptive model is developed through subjective responses of the occupants and experimental measurements.

To evaluate the natural airflow in the classrooms is used tracer gas decreasing concentration, using the carbon dioxide released by the occupants.

The classrooms are ventilated by natural airflow, the greenhouse (hall) is ventilated by forced airflow and the playground is ventilated by forced airflow also.

In winter thermal conditions the forced airflow is done from the greenhouse to the indoor playground, while in summer thermal conditions the forced airflow is done in the opposite direction.

4.7. Cross Ventilation in a School Building

In Conceição et al. (2012b) the evaluation of indoor air quality in a classroom equipped with crossed ventilation is made. The experimental study is made in a school building, located in the South of Portugal, and the numerical study is made by the building thermal response software.

The study is divided in two parts: in the first one the air exchange rate with different airflow typologies is made in different classrooms, equipped with natural and cross-flow ventilation, using the tracer gas decay method (using carbon dioxide concentration) inside spaces, while in the second one the calculated indoor air quality is used in the numerical simulation to evaluate the indoor air quality inside the spaces.

4.8. Mixing Ventilation and Cold Radiant Floor

In Conceição et al. (2012c) the comfort and airflow evaluation in spaces equipped with mixing ventilation and cold radiant floor is made. The numerical study is made in a virtual chamber, with $2.7 \times 2.4 \times 2.4$ m^3, using the coupling of differential computational fluids dynamics and integral human thermo-regulation software.

In this study is used the predicted mean vote (PMV) index, calculated by the integral model. The ventilation effectiveness is obtained by the differential model. An air distribution index (ADI) for the occupied spaces with non-uniform environments is developed.

The airflow inside a virtual chamber equipped with two occupants seated in a classroom desk, is promoted by a mixing ventilation system with supply air of 28°C and by a cold radiant floor with a surface temperature of 19°C. The mechanical mixing ventilation system uses a supply and exhaust diffusers located above the head level on adjacent walls.

References

E. Z. E. Conceição and Mª M. J. R. Lúcio "Evaluation of Thermal Comfort Conditions in a Localized Radiant System Placed in Front and Behind two Students Seated Nearby Warmed Curtains", *Building and Environment*, Volume 45, Issue 10, October 2010, pp. 2100-2110.

E. Z. E. Conceição and Mª M. J. R. Lúcio "Evaluation of Thermal Comfort Conditions in a Classroom Equipped With Cooling Radiant Systems and Subjected to Uniform Convective Environment", *Applied Mathematical Modelling*, Issue 3, Volume 35, March 2011, pp. 1292-1305.

E. Z. E. Conceição, J. M. M. Gomes, N. H. Antão, and Mª M. J. R. Lúcio "Application of a Developed Adaptive Model in the Evaluation of Thermal Comfort in Ventilated Kindergarten Occupied Spaces", *Building and Environment*, Volume 50, April 2012a, pp. 190-201.

E. Z. E. Conceição, J. P. Farinho and Mª M. J. R. Lúcio "Evaluation of Indoor Air Quality in a Classroom Equipped with Crossed Ventilation", *The International Journal of Ventilation,* UK, June 2012b, Vol. 11, No. 1, June 2012b, pp. 53-67.

E. Z. E. Conceição, Mª M. J. R. Lúcio and H. B. Awbi "Comfort and airflow evaluation in spaces equipped with mixing ventilation and cold radiant floor", *Building Simulation*, September 2012c, pp. 1-17.

E. Z. E. Conceição, Mª M. J. R. Lúcio, S. P. Rosa, A. L. V. Custódio, R. L. Andrade, and Mª J. P. A. Meira "Evaluation of comfort level in desks equipped with two personalized ventilation systems in slightly warm environments", *Building and Environment*, Volume 45, Issue 3, March 2010, pp. 601-609.

E. Z. E. Conceição, Mª M. J. R. Lúcio, V. D. S. R. Vicente, and V. C. T. Rosão "Evaluation of Local Thermal Discomfort in a Classroom Equipped with Crossed Ventilation", *The International Journal of Ventilation*, UK, Vol. 7, Nº 3, December 2008b, pp. 267-277.

E. Z. E. Conceição, V. D. S. R. Vicente and Mª M. J. R. Lúcio "Airflow Inside School Buildings Office Compartments With Moderate Environment", *International Journal on Heating Air Conditioning and Refrigerating Research,* ASHRAE, American Society of Heating, Refrigerating and Air-Conditioning Engineers, Inc., EUA, Vol. 14, Nº 2, March 2008a, pp. 195-207.

INDEX

#

A

B

C

D

I

K

L

M

N

O

P

Q

R

S

T